ASSASSIN BUGS, WATERSCORPIONS, AND OTHER HEMIPTERA

REPRODUCTIVE BIOLOGY AND LABORATORY CULTURE METHODS

Orin McMonigle

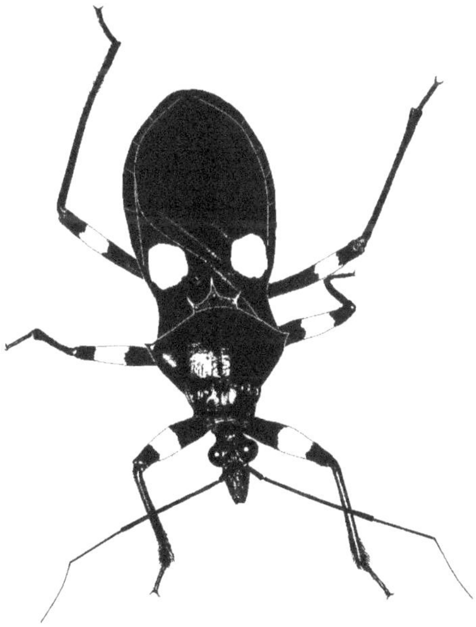

COACHWHIP PUBLICATIONS
GREENVILLE, OHIO

DEDICATION

Dedicated to my sisters Megan and Erin, and my brother Ryan who spent countless summer days with me hunting for bugs.

ACKNOWLEDGMENTS

I would like to thank Christian Schwarz, Annie Lastar, Gillian Higgins, and Cameron Campbell for editorial assistance. I would also like to thank Randy Morgan, Peter Grabowitz, Mike Pickut, Don Ahart, Dexter Davson, Joe Murphy, Matt Smith, Alan Jeon, Jordan Landerman, Jim Smolka, Stephanie McDorman, and Steve Franz for providing information or specimens for photography. I appreciate the opportunity I had to review the Heteroptera collection of the Cleveland Museum of Natural History. The wonderful exhibits at the Cincinnati Zoo's Insect World allowed for observation of a number of live true bugs and husbandry considerations. The encouragement and input from my wife Sylvia was greatly appreciated. Finally, my greatest thanks to God for creating so many neat bugs.

Assassins, Waterscorpions, and Other Hemiptera

ISBN 1-61646-426-7
ISBN-13 978-1-61646-426-4

CoachwhipBooks.com

CONTENTS

Wheel bug nymphs and eggs on a twig in
south-central Pennsylvania. (cc) Gordon Brinton Jr.

INTRODUCTION

THE LARGE SIZE and impressive appearance of certain true bugs makes them sought after for invertebrate displays and useful for classroom study and scientific investigation. Such a varied order includes numerous species that attract the entomologist's attention as well as the layperson's curiosity. The commonly kept giant water bugs and waterscorpions are fascinating predators of fish, tadpoles, and other aquatic life. Easily propagated shield bugs and seed bugs, as well as the more demanding leaf-footed bugs, are exquisitely decorated and brightly pigmented in an endless array of colors. One colorful seed bug, the large milkweed bug (*Oncopeltus fasciatus*), has seen regular usage in classroom research for decades and has shown up in various insect zoo exhibits. Legions of assassin bugs are impressively ornamented with bright pigments and huge spines or ridges. They feed ravenously on pest insects, and yet are seldom aggressive towards their keepers. Some of the hardy African *Platymeris* stocks have been in continuous culture for at least twenty-five years (McMonigle 2001), with at least one species being reared in numbers in the laboratory at least as far back as 1961 (Manjanuth et al. 1969) for biological pest control. Even an improbably large species of blood-feeding bug, the Baja giant conenose (*Dipetalogaster maximus*) has been a popular display animal in at least one insect zoo for more than two decades (per the Cincinnati Zoo stock list of 1993). This text is dedicated to furthering the understanding of reproductive biology and advancing captive propagation documentation for this spectacular order of animals.

The heteropteran life cycle is often well-suited for simple and homogeneous captive propagation which lends itself to research and school projects. The eggs of most species are quite sturdy and hold up well to handling and some lack of attention. Nymphs grow quickly and are seldom prone to bad molts. Adults and nymphs eat the same foods and require the same or similar care (unlike most lepidopterans and coleopterans). Few species require

Laccotrephes calcar first instar nymph.

more than three months to reach adulthood and the adults of most temperate species can be held in diapause for long periods. The life cycle can usually be observed from start to finish in the classroom over less than a semester and experiments requiring multiple generations are easily achieved. Quick turnaround allows researchers to switch some bug species to more readily available foods and select for desirable traits (so far those traits have been reserved to survivability in captivity rather than size or color as seen in cultivated isopods, shrimp, cockroaches, etc.). Most adult bugs are long-lived and survive from one to four years (conversely, most adult moths, butterflies, beetles, grasshoppers, mantids, etc. live just a few months after reaching maturity).

Nymphs and adults of the popular lab animal, *Oncopeltus fasciatus.*

There has been very little in the way of literature related to laboratory methods for the captive reproduction of true bugs. A handful of short articles referencing aquatic bug rearing have shown up in periodical literature (Morgan 1988, 1996, Morgan and Thoms 1994).

Shield Bug

Biological supply houses have offered care sheets on large milkweed bugs since the late 1980s. A few general husbandry texts have included a page or two, such as the slightly over one page (out of 320) on *Platymeris* spp. in Bruins (1999), two pages on keeping aquatic hemipterans in Dunn (1993), or coverage of both groups over five pages in McMonigle (2011). A few German-language general terrarium texts have also included a page or two on *Platymeris* culture (Kleinsteuber 1989, Löser 1991). In periodical literature there have been some articles on culturing *Platymeris* by Eluwa (1972), Larsen (1998, in Danish), Fritzsche (2003, in German), and Yelich (2004), two on *Euthyrhynchus* (McMonigle 2014, Murphy 2014), one each on *Arilus* (Hagerty and McPherson 2000) and *Emesaya* (Hagerty et al. 2001), and a few on laboratory requirements of *Abedus* (McMonigle 2001b, 2004b) and *Laccotrephes* (McMonigle 2004a). At the date of writing the only dedicated true bug husbandry text ever composed was *Assassins, Waterscorpions & Other True Bugs: Care and*

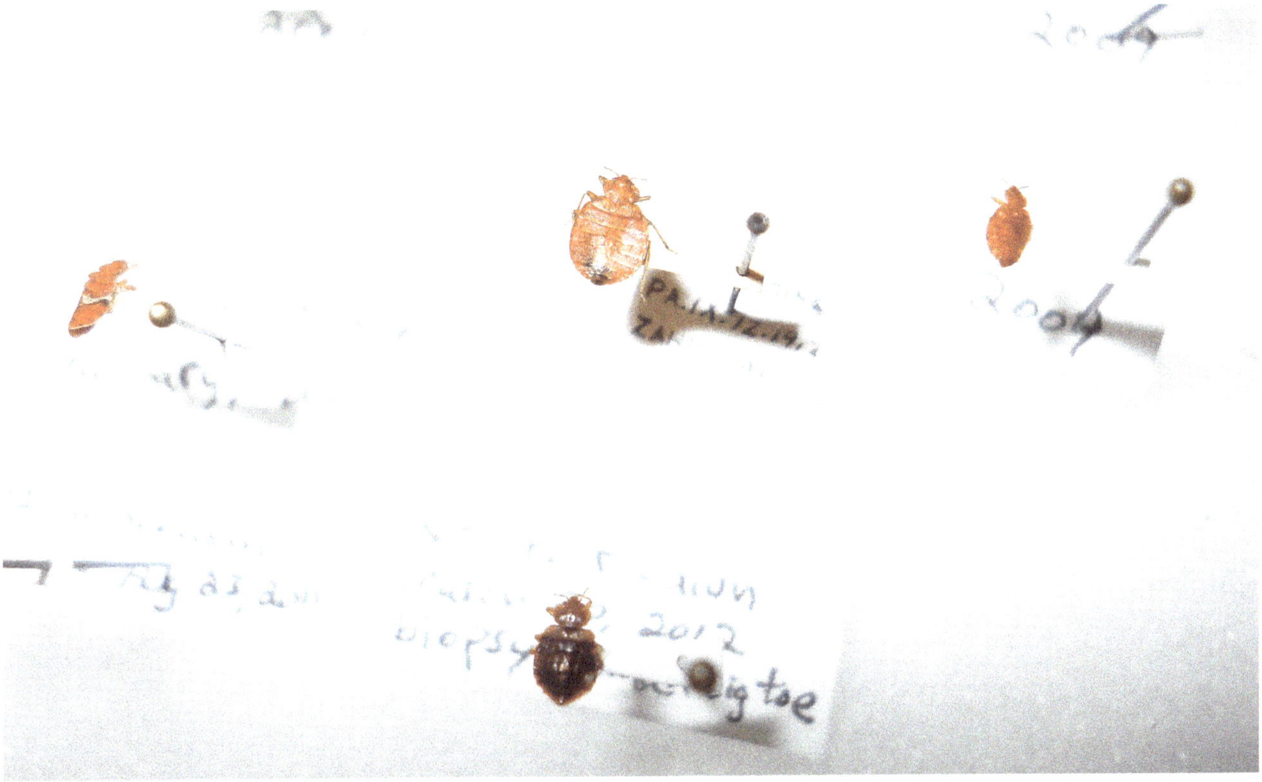

Bed bugs may be the most famous of the true bugs.

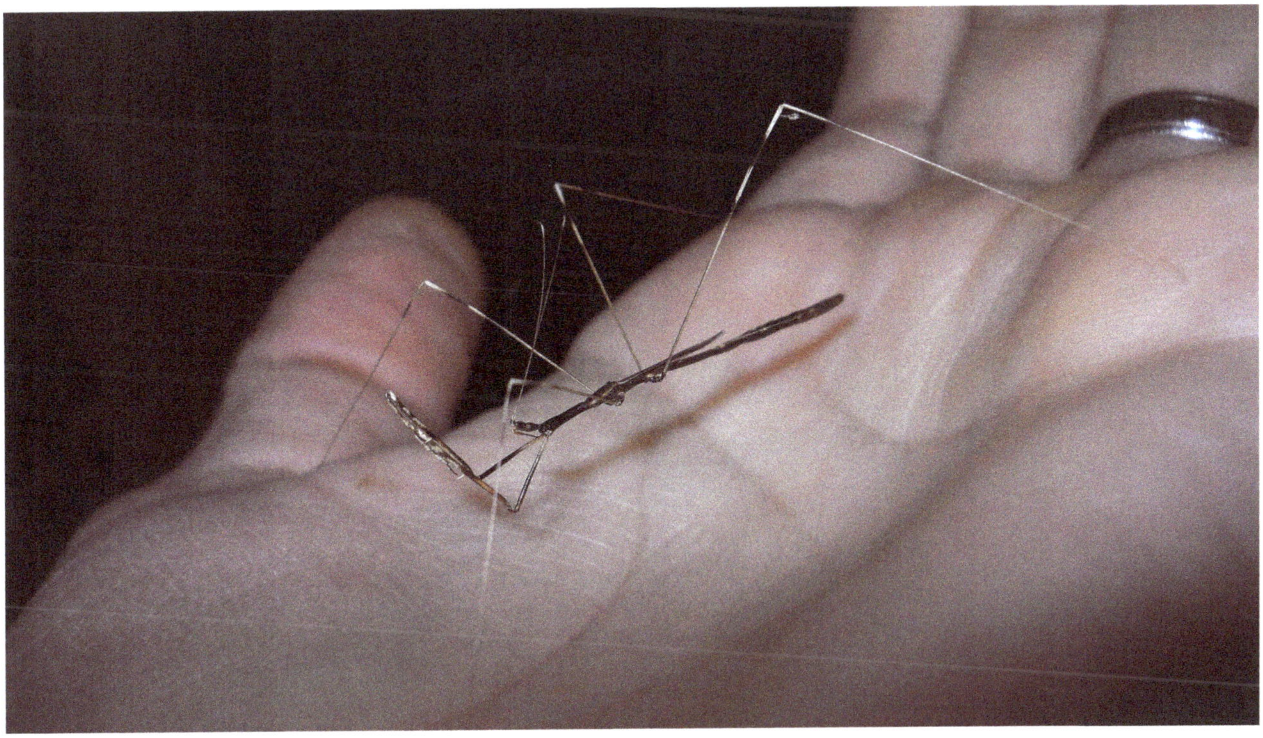

The amazingly long legs of *Emesaya brevipennis*. Specimen from Girard, Ohio.

Some predatory shield bugs are reared for their ability to control pests.

Culture (McMonigle 2001a). Though a small booklet of thirty-six pages, it contained culture methods previously unpublished. Some stocks like the large milkweed bug, the ferocious water bug (*Abedus herberti*), and certain African assassin bugs (*Platymeris* spp.) had been kept in captivity for a decade or more before that text was published, yet with little basic husbandry documentation.

This book expands on *Assassins, Waterscorpions, & Other True Bugs*, including information from various true bug *Invertebrates-Magazine* articles, and current experience. It is written to provide information on keeping and feeding the most commonly propagated true bugs. Each chapter includes information on background and classification, dietary requirements, life cycle, and housing for the families discussed. Life cycle information presented throughout the chapters is predominantly based on laboratory observations under artificial lighting (14-16 hr day / 10-8 hr night), at room temperature (72-75° F, 22-24° C), with food or prey items different from those found in nature. Results are not likely to be identical to any wild situation.

Unless the reader lives at the North or South Pole, in the middle of a desert, or on an oceanic platform, members of every family of true bug discussed here will be easy to collect. In Europe there is greater access to wild, tropical specimens from vendors, but native species

Camouflaged assassin nymph, *Lisarda* sp., Panay Island, Philippines. © Christian Schwarz

are more likely to be protected, even when they are common and at no risk of disappearing. In North America importation is strictly regulated (even for documented beneficial predators) but native specimens are nearly always legal to keep. In Hawaii, all the native species are strictly protected (because their populations have been decimated by habitat loss and purposeful agricultural introductions) while introduced animals are fair game. Specific regulations for one's city, state, or country should be researched before collecting or transporting livestock. Depending on location, a handful of species (at most) may be purchased through biological supply houses or invertebrate vendors.

There are various reasons for writing a massive, time-consuming text for a group whose captive propagation has always been limited. Since the 2001 *Care and Culture* text there has been significant improvement in knowledge. The original book had very little room for expansion because of its format. However, this format is relatively unlimited, providing an opportunity to include all the current details and to thoroughly describe husbandry experiences. Most of all, the advent of an amazing and unbelievable new species prompted this effort.

My experience in laboratory culture of true bugs stretches back more than two decades. This includes 13 years of continuous culture of

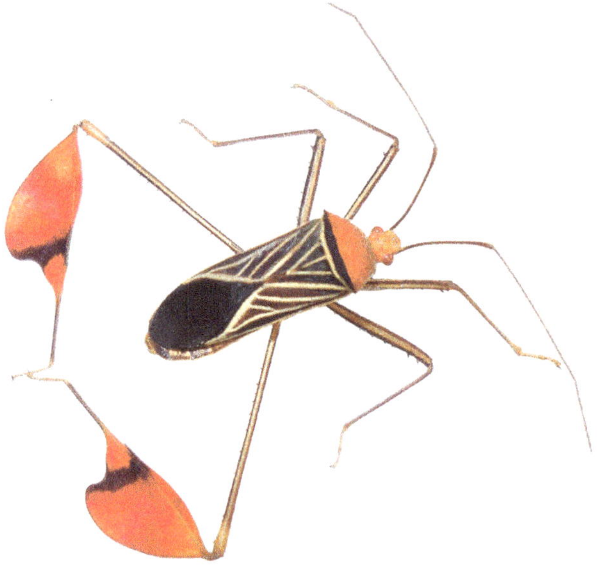

Leaf-Footed Bug

a single-bloodline Tanzanian *Laccotrephes* stock, 10 years of continuous culture of a single-bloodline *Abedus* stock, shorter stints with other *Abedus* stock, decades with three species of *Platymeris*, as well as dozens of full species cycles or multiple generations with various true bug families. There are few who can offer such extensive experience. Hopefully I can translate this into useful explanations and reproducible guidelines. Husbandry information is based on methods that have been successfully employed to provide a recommended starting point for a new keeper. You may discover more useful methods, foods, or enclosures with time and experience.

An attempt has been made to include as many types of true bugs as is reasonably possible and to offer details for species found on different continents. Nevertheless, captive propagation for a *Ranatra* sp., or *Lygaeus* sp., whether it hails from Africa, Asia, Australia, or the Americas, is going to be essentially the same. Species detailed here are, of necessity, primarily those either collected or commonly reared in culture in the U.S., and the focus is on non-pest species. While taxonomy and classification are important to further the understanding of true bugs, and much effort has been made to conform to the most recent considerations, this text is not intended to further any taxonomic goals. U.S. species numbers in the chapters are according to Arnett (2000). Worldwide counts are adapted from Henry (2009).

Large nymphs of a magnificent species new to laboratory stocks.

Red spiny assassin bug (*Scipinia arenacea*) eating a ladybug. (cc) James Niland

Ultimate molt and the acquisition of wings.

Bugs, like any arthropod, are subject to bad molts resulting in damage and death.

Large numbers of *Platymeris* sp. “Mombo” in 2004 (1999 stock).

Abedus herberti males with eggs on their backs at a local bug exhibit.

THE TRUE BUG

Although most people lump everything from spiders and butterflies to germs and computer glitches into the "bug" category, there is an order of insects technically named the "true" bugs. Members of the order Hemiptera, suborder Heteroptera, the true bugs, are distinguished from other insects by the possession of sucking mouthparts located at the front of the head, two pairs of wings, an incomplete life cycle, and nearly always a thickened front half of the outer wings. The name of the order (*hemi* means "half" and *pteron* means "wing") refers to the form of the heteropteran forewing. The hemipterans comprise a huge order containing 82,000 described species, more than any other hemimetabolan order. 42,000 of the described species fall within the Suborder Heteroptera (Henry 2009).

Despite their prevalence in most ecosystems and their tendency to be relatively large and colorful in relation to other terrestrial invertebrates, heteropterans are hardly well known to the general public. The vast majority of people who run across a giant water bug say they had never seen such a big cockroach before. The general public is familiar with water striders, though many of them confuse *strider* with *spider*. True bugs are not among the feeder insects found at pet shops, nor are they among the common invertebrate pets. Entomology books other than the very rare text dedicated to true bugs almost never see a bug on the cover. (One exception is the 2007 Kaufman field guide with a large milkweed bug featured squarely on the front.) Assassin bugs have gained fame in popular television, starring in episodes of *Monster Bug Wars* (Series 1, Parts 3 and 6, 2011, and Series 2, Parts 3 and 4, 2012). Stink bugs or shield bugs are probably the most familiar member of the group, though they are confused with any insect that has a defensive odor like the *Eleodes* beetles. The big, colorful, tropical stink bugs make their

Camellia shield bug *Poecilocoris latus* glow in the dark necklace, a tessaratomid nymph bracelet from China, and a man-faced bug *Catacanthus incarnatus* paper weight.

way to paper weights and necklaces with some regularity.

This book details representative members of those heteropteran families attractive to the typical researcher, entomologist, enthusiast, or classroom teacher. Some of these species may hold promise for development of cancer treatment or other medicines through the studies of their venom or defensive secretions. Others have been investigated and employed for their ability to target pests. These include assassin bugs to control pestiferous beetles and giant water bugs to control disease vectors like mosquito larvae and aquatic snails through predation. Even the herbivorous leaf-footed and seed bugs have been tested for their utility in the control of specific noxious weeds. The vertebrate pest families of the bed bugs (Cimicidae) and bat bugs (Polyctenidae) are mentioned briefly only for their novelty. Even the most dedicated researcher would be unlikely to have the funds and ability to rear bats or birds to recreate the habitat of members of these two families (rather than simply study them in the field). Though blood feeders can usually go for a very long time without food, it may be too difficult to substitute their hosts through generations. Few individuals would consider purposefully keeping alive, let alone offering their blood to pet bed bugs (*Cimex lectularius*) even though: ". . . extensive testing has determined that bed bugs are incapable of transmitting human pathogens" (Cranshaw et al. 2013). There is already a wealth of research on *C. lectularius* and *Triatoma infestans*. Blood feeding members of the Reduviidae (subfamily Triatominae) are mentioned because of their commonly kept relatives. However, unlike insectivorous assassin bugs, the blood-sucking Reduviidae do not make good culture items since they will suck human blood if they escape and are capable of transmitting vertebrate disease, primarily Chagas disease in Central and South America. Nevertheless, Triatominae in North America are unlikely to find a source for the disease and most Triatominae are not vectors since they are not adapted to live with humans or they do not excrete waste on their host. The vast majority of Reduviidae feed on invertebrates and cannot survive on vertebrate blood. Most bug families not

Platymeris biguttatus, this captive stock dates back to 1994, possibly the 1970s or 1980s.

Assassin bugs especially, can be very colorful when teneral.

A plant bug (Family Miridae) from Argentina. Plant bugs are often very handsome when magnified greatly. © Martin Gamache

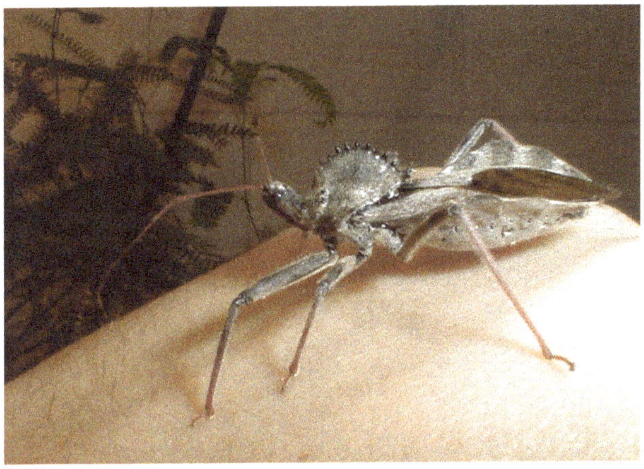

Arilus cristatus can inflict a painful bite, but does not recognize flesh as food.

Sticky front legs used for capturing prey, reduviid from Suriname. © Martin Gamache

Flat bugs (Family Aradidae) are small bugs usually associated with tree bark. Aradids are found in temperate climates, but are often far less ornate than this tropical *Dysodius* sp. photographed in French Guiana. © Martin Gamache

discussed in the following chapters require a magnifying lens for observation. The tiny members of the Nabidae, Aradidae, Miridae, Tingidae, Hebridae, Naucoridae, and numerous other true bug families are rarely maintained in captivity and are interesting only to the more zealous heteropteran enthusiasts or those studying specific vectors.

While Hemiptera means "half-wing" because it refers to the divided membranous and thickened outer wings seen only on the heteropterans, the remaining subfamilies have generally homologous wings when present. The "old" Homoptera (currently two different suborders of the Hemiptera) includes a wide range of plant feeding species including the fantastically large and loud cicadas, the incredible lanternflies, pestiferous aphids, fungus-like scale insects, and small, but ornate leaf and treehoppers, among others. These, mostly, are restricted to a life sucking the plant juices from living plants making larger species like plant hoppers and cicadas difficult if not impossible to keep in readily maintained captive conditions (although pestiferous aphids and scale insects are easy to produce in quantity and can even thrive on cuttings). Keeping the host plant alive for long periods while it is being fed upon, though difficult and resource consuming, can be the easiest part. Egg laying, long development and gestation, or short adult lifespan can present uniquely difficult challenges. The only member of this group I have reared through generations (not including unwanted aphid and scale infestations) was the buffalo treehopper (*Stictocephala* sp.). I maintained these on miniature roses in a screen-topped, glass terrarium. The roses held up for years despite acquiring a very knobby appearance from feeding. Although successful in terms of generations, the colony production never exceeded a few dozen adults. This revised subset of the Hemiptera is not covered in this book because they are not true bugs and species other than tiny pests are not suited to laboratory culture.

Buffalo Treehopper

True bugs have adapted to a variety of niches. Some feed on plants and molds, some are predatory, and others are scavengers, detritivores, or omnivores—some even exist as parasites. They live in such diverse habitats as forests, deserts, grasslands, human dwellings, animal dens, bat fur, bird feathers, tundra, puddles, ponds, marshes, lakes, rivers, and the water's surface. They are the only insects to have adapted to life on the open ocean. They are distinctly absent from the depths of lakes and below the surface of the sea. Seed, leaf-footed, shield, and assassin bugs are found throughout most terrestrial environments. Bloodsucking conenoses and bed bugs live near their human or animal prey while certain assassin bugs specialize on eating these same bloodsuckers. An entire host of species in two families lives in the fur of bats, ingesting their blood. Giant water bugs, backswimmers, water boatmen, and waterscorpions inhabit the largest and smallest bodies of fresh water. Water striders can be found on the tenuous surface film of nearly every body of water on earth.

The true success of this group may relate to their ability to grow quickly and modify behavior, or it may result from the design of the body. Protective outer wings offer adaptability and are mirrored in some of the most successful insect orders like beetles and orthopterans, though hymenopterans and lepidopterans are also widespread and myriad without such mechanical protection. True bug body parts tend to be relatively standard in design for an insect, even if some are highly adaptable or unique.

WINGS

Hemelytra comes from the prefix *hemi-* meaning "half" and *elytra*, the name for the hardened wingcases of beetles. These and the sometimes massive scutellum are designed to shield and protect the membranous hind wings since they are easily damaged. Wings are significant in dispersal to new areas and movement within

Most true bugs have functional wings hidden beneath the hemelytra.

Rear leg adaptations. *Leptoglossus phyllopus.*

populations, but despite this many species have given up their wings. Wings also provide the ability to escape predators through flight, and they can be used by some species in a startle display. A number of species across various families (mostly shield bugs and assassin bugs) ruffle their wings to make noises when disturbed. The upper surface of the abdomen of certain large bugs is colored in crimson red. When disturbed, they open their wings to flash a red warning sign; but wing coverage may also aid in avoiding predatory birds since the bright color disappears immediately when the bug lands.

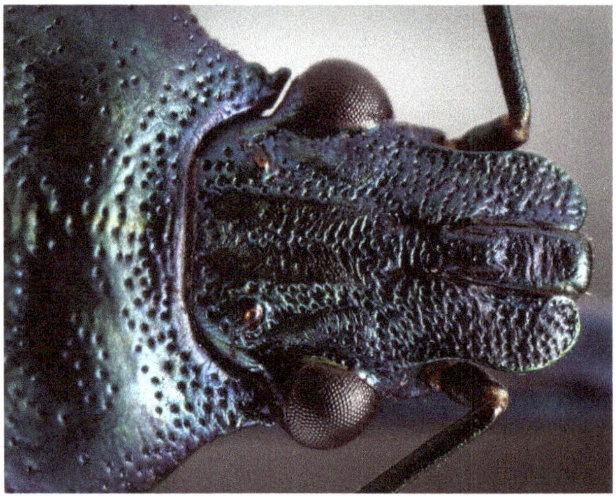

Note the two ocelli behind and between the compound eyes. © Henry Kohler

LEGS

The legs are often of standard morphology, but they range greatly across the order. Predatory species commonly have curved, inflated, or elongated front legs for the capture of prey. Front legs can be spine-covered, hairy, coated in a sticky secretion, plate-like, or raptorial and sharp like the front legs of a mantis. Some herbivores (notably the leaf-footed bugs and male giant shield bugs) have legs that are expanded into massive, strong femora with thick inner spines or flattened and shaped like vegetation. These rear leg adaptations range in purpose from battling over mates to courtship or crypsis. Species across various groups have one or more pair of legs that are extremely long for one purpose or another. Aquatic bugs can have legs that are elongate and flattened into paddles.

EYES

The primary pair of large eyes is of compound design and offers excellent vision for both predators and herbivores. The individual facets are too small for the naked eye to see and commonly range from ~500 to 2,000 per eye for most terrestrial species. Giant water bugs can have 4,000 facets in a single eye. In addition to the pair of large compound eyes, many heteropteran families and subfamilies are outfitted with two notable ocelli located between and behind the compound eyes. Most other insects that possess ocelli (like wasps, mantids, and cicadas) have a cluster of three. Each ocellus has a single lens that can detect dark, light, and color rather than movement.

ROSTRUM

The rostrum, designed to pierce objects for feeding, is variable in design across families and species but is always located towards the front of the head and is relatively prehensile. (Aphids, planthoppers, etc. have the rostrum located near the base of the head, while the labium generally does not have internal musculature and bendable joints as seen in Heteroptera.) The central stylet is composed of four segments (two maxillae and two mandibles) which combined are rarely much wider in cross section than a human hair. A far bulkier structure, the labium, surrounds the stylet and acts primarily as a fulcrum against which the stylet pushes to drill into food. The stylet is highly dexterous and whip-like, while the labium is

Assassin bug *Sirthenea carinata* face, Maryland.
(cc) Sam Droege / USGS Native Bee Inventory and Monitoring Program

Indonesian assassin bug face. (cc) Donald Jusa

Many true bugs have a long, bendable rostrum like this *Euthyrhynchus floridanus* from Alabama. © Henry Kohler

The rostrum is powerful enough to hold massive prey by an improbably small point because the stylet provides an anchor inside the prey.

Front leg adaptations. *Euthyrhynchus floridanus* from Alabama. © Henry Kohler.

Front leg adaptations. *Phymata* sp., Pennsylvania. © Henry Kohler.

muscular and can be moved in different directions during feeding.

SETAE

These structures are the most numerous feature and may be the most important single adaptation of the true bug body. They appear similar to animal hairs but are entirely different structures formed from the material of the exoskeleton. They are most often adapted as sensory organs that transmit information from the environment, but also act as coatings to protect from desiccation, coverings to aid in camouflage, fringes for propulsion in swimming, and aquatic lungs for underwater breathing.

Setae are important sensory organs that cover the heteropteran body and provide information from the outside world.

Arilus cristatus setae. Specimen from Bethlehem, Pennsylvania. © Henry Kohler

Setae can be used to hold air bubbles or function
to absorb oxygen directly from the water.

GIANT WATER BUGS AND WATERSCORPIONS (NEPOIDEA)

The well-known giant water bugs (family Belostomatidae) and waterscorpions (family Nepidae) are closely related families in the Superfamily Nepoidea. Both are equipped with a breathing tube (composed of two segments) located at the end of the abdomen and both live just below the water's surface. In addition, both are semi-aquatic predators armed with a short rostrum and raptorial front legs adapted for prey capture. The rear and middle legs are used in locomotion and tend to look very different from the forelegs. All species have large compound eyes used to locate prey and small antennae tucked in a groove beneath the eyes. The body coloration (nearly always some variation of brown) and the body design are relatively uniform. Related species found in various parts of the world appear nearly identical. For example, there is at least one species of *Lethocerus* on every continent (except Antarctica) resembling the North American *Lethocerus americanus*, while *Nepa* spp. in the Americas, China, Europe, and Asia can be difficult to tell apart without dissection of the male genitalia.

Lethocerus medius from southern Arizona. Milky bluish-white defensive secretion is visible on the head, above the rostrum.

Although certain types of waterscorpions and giant water bugs appear to be similar at a glance, there are two visible differences between the two families:

1. The breathing tube of Belostomatidae appears short and can be fully retracted. Rarely more than an eighth of an inch (0.2 cm–0.3 cm) if any, can be seen sticking out past the abdomen or wings of a live specimen. In contrast, the non-retractable breathing tube of the Nepidae is entirely visible and ranges from half an inch to two and a half inches (1.2 cm–6.4 cm) in length.

2. Giant water bugs usually have thick, paddle-shaped middle and hind legs adapted for swimming. For waterscorpions, the same legs are thin and usable only for walking and hanging onto underwater objects. They may or may not have thick fringes of setae to aid with limited swimming.

GIANT WATER BUGS (BELOSTOMATIDAE)

This family contains the most massive true bugs. Despite rumors, the largest water bug in the world does not exceed five inches (< 12.5 cm) in length. (See attached image: 1. *Lethocerus maximus* French Guiana. 2. *Kirkaldyia deyrolli* Wangnuea Lampang, Thailand. 3. *Lethocerus medius* Madera Canyon, Santa Cruz County, Arizona. 4. *Benacus griseus* Danslow County, North Carolina.) Even the small species are large by insect standards. The creeping water bugs in the Family Naucoridae look similar and could be confused with Belostomatidae, but are smaller (usually around 1 cm) and uncommon in temperate areas. Giant water

These Lethocerinae may not seem impressive if you do not realize the third one from the left is one of the largest insects in North America (*Lethocerus medius*). The far more massive creature on the left is from South America. Left to right specimens are: 1. *Lethocerus maximus*, French Guiana. 2. *Kirkaldyia deyrolli*, Wangnuea Lampang, Thailand. 3. *Lethocerus medius*, Madera Canyon, Santa Cruz County, Arizona. 4. *Benacus griseus*, Danslow County, North Carolina.

bugs are massive and amazing creatures, exhibiting parental care of eggs. Although they are common from the tropics to cold temperate areas, few people are familiar with them. The term "water bug" is more commonly recognized as a description of pest cockroaches (primarily *Blatta orientalis*) because they live in damp areas of basements. Belostomatids' only claim to fame is their popularity as an Asian delicacy.

Within the Family Belostomatidae there are two primary subdivisions: large (Subfamily Lethocerinae) and small (Subfamily Belostomatinae) giant water bugs. The principal differences between large and small Belostomatids are where the eggs are laid and, as the names denote, physical size. Belostomatinae do not exceed 1.6″ (4 cm) while Lethocerinae are normally greater than 1.6″ (4 cm) in length. Furthermore, small giant water bugs tend to have a more rounded, broader body shape, narrower middle and hind legs, and a longer or thinner rostrum. Belostomatinae genera include *Abedus, Belostoma, Diplonychus, Limnogeton,* and *Sphaerodema*. The ferocious water bug *Abedus herberti,* found from Mexico to the southwestern U.S., can exceed an inch and a half in length (> 3.5 cm) and is among the most massive of all the small giant water bugs. Most Belostomatinae are an inch or less (< 2.5 cm). Within the Lethocerinae, there are two dozen or so *Lethocerus* species but only one species each in the genera *Benacus* and *Kirkaldyia*. *Lethocerus americanus* is representative of the large giant water bugs. It is one of the largest North American insects and ranges from two to two and a half inches (5-6.5 cm) in length, while *Lethocerus maximus* from northern South America is the world's largest species at approximately five inches (12.5 cm).

Two nearly mature *Hydrachna* species parasitic mites are attached near the base of the hemelytra (left) and three protonymphs (1st instar mites) can be seen on the right; wild specimen *Belostoma flumineum*.

Giant water bugs inhabit streams, rivers, puddles, and medium-sized pools in addition to larger bodies of fresh water. They prefer minimal water movement and so are only encountered in calm areas of streams and rivers. *Abedus herberti* can be found in the pools of fast moving mountain streams. *Belostoma* and *Lethocerus* are often found in weedy areas of ponds and lakes. *Belostoma* are found among small-leafed floating pondweeds while *Lethocerus* are associated with larger plants such as water lilies. Horse troughs and water-filled storage tanks are common manmade homes for giant water bugs. Many of the smallest species have adapted to life in temporary pools and adults easily fly away when their puddles dry up. Adult Belostomatidae are strong fliers and are attracted to lights after dusk. In many areas

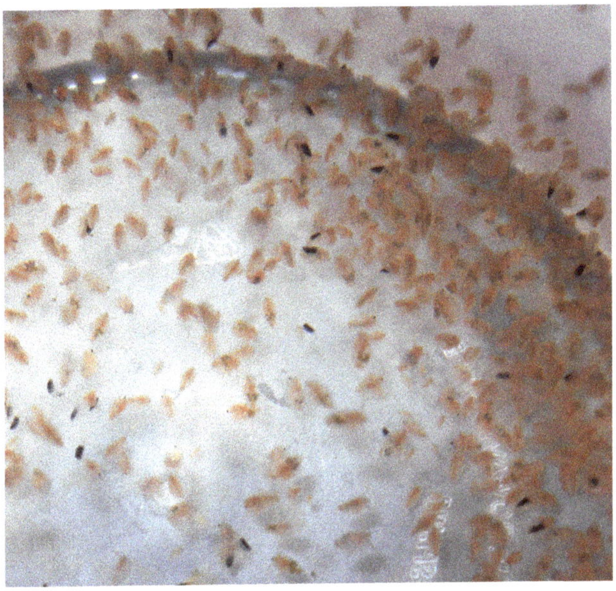

Daphnia magna is a nutritional item for feeding 1st and 2nd instar giant water bug nymphs. The black specks are resting eggs. Females give birth to live young but can produce resting eggs. Under adverse conditions reproduction is only resting eggs. This aquatic prey can be cultured in vats of fresh water and fed frozen green peas (*Pisum sativum*). The peas are mixed with two parts water (a multivitamin can be added) in a blender and liquefied. A few tablespoons are added to each vat every few days as consumed.

Common aquarium snails useful for feeding *Belostoma*: *Physa gyrina* (top), *Planorbarius corneus* (left), and *Melanoides tuberculata* (bottom right).

they can be seasonally collected in huge numbers under the giant bluish-white lights of gas stations and shopping malls. Twenty of the 150 or so belostomatid species are found in the U.S.

DIETARY REQUIREMENTS

Giant water bugs are ravenous predators who actively pursue and attack prey. As well as hunting in water, most will jump out of the water to grab small creatures at the water's edge. Within a fraction of a second, they pierce their victim with their thin stylet and inject a cocktail of digestive fluids and poison. Belostomatids are formidable predators that feed on vertebrates, specifically fish and tadpoles, in nature. The largest species have been recorded feeding on more massive vertebrates such as salamanders, small snakes, and turtles. They pose little threat to even the smallest of mammals, unless those mammals find themselves in the water. There is an old report of a woodpecker being fed upon (Hutchins 1966). Giant water bugs are usually considered beneficial because their primary prey includes disease vectors, specifically mosquitoes and aquatic gastropods. Larger nymphs can eat a hundred or more mosquito larvae before reaching the next instar. The strong muscular foot of aquatic snails does not prevent them from being pried loose and feasted upon by giant water bugs. Any common aquarium snails can be used as prey though one of the most common, the Malaysian trumpet snail (*Melanoides tuberculata*), presents a physical challenge. This snail has a shell lid (operculum) and a long spiral shell, but it is often consumed if the bug is hungry enough. African Belostomatids in the genus *Limnogeton* have less pronounced raptorial forelegs because they are specialized

Predation photographed in nature. (Top) *Kirkaldyia deyrolli* feeding on a snake.
(Bottom) *Kirkaldyia deyrolli* eating a Reeve`s pond turtle in a ditch next to a rice field.
© Shin-ya Ohba (Nagasaki University, Japan)

in snails and do not bother catching fast moving aquatic invertebrates or fish. Some North American species of *Belostoma* behave similarly.

There are a number of fascinating aspects of Belostomatid prey capture and feeding that can be observed in the laboratory. *Abedus herberti* nymphs become accustomed to feeding and can learn to grab foods thrown into their tanks before it touches the water. Whether they become familiar with their caretaker's movement or become used to the airborne food's trajectory is uncertain. On the other hand, many species require patience when trying to view feeding habits because they hide when their keeper approaches. Unlike most predators, which drop their food and run if disturbed, giant water bugs tend to hold their food tight and even take it with them deep underwater to hide. Possibly the most fascinating and unsettling aspect of Belostomatid feeding is only noticed when they are given food with a see-through carapace such as a beetle pupa. Projecting from the short rostrum is the incredibly dexterous, long, thin stylet that can be seen scraping the insides of the furthest sections of the food animal (reminiscent of the extremely long and highly dexterous velvet ant stinger, except the stylet is never seen flicking into the air). This is why *Belostoma* is capable of completely scraping out the contents of a spiral snail shell whose recesses are far beyond the reach of the measured length of the rostrum.

Giant water bugs accept a variety of aquatic and terrestrial, live or dead foods. Small fish (guppies or feeder goldfish) and crustaceans (daphnia or brine shrimp) are well-accepted and readily available aquatic prey items commonly reared for use in feeding aquarium fish. Cultured *Daphnia magna* offers the best chances for growth and survival of early immatures, but the quantity required to rear a few water bugs beyond the first two instars would require large rearing vats for the daphnia. Cultured flatworms are sometimes found and eaten by small *Abedus* nymphs, but otherwise are only good as clean-up crews. Aquarium snails, including the common ramshorn *Planorbarius corneus* and pond snails such as *Physa gyrina*, are useful prey for less aggressive *Belostoma* species, but can also be eaten by hungry *Abedus* and *Lethocerus*. Belostomatids from 1st instar nymph to adult will feast on terrestrial prey thrown on the surface. Mealworms, crickets, isopods, and cockroaches are readily purchased or cultured to feed large nymphs and adults. Freeze-dried crickets or mealworms can be tossed on the surface because giant water bugs do not require live prey. Floating fish food pellets (fish or shrimp meal based, not plant based) can also be a satisfactory food. Some scavenging is common, but items that are not immediately grabbed and fed on will simply rot. Live prey which drowns and sinks to the bottom is much more likely to be ignored than floating items.

Hatchlings must be fed within the first three days or they die. Any size land arthropod can be tossed in for the hatchlings' first meal. Readily available prey items include fruit flies and pea weevils for small nymphs. Although tiny nymphs cannot tackle live food items more than three to four times their size, they can feed on incredibly massive prey when it drowns and stops moving. However, sunken, motionless prey is usually ignored and rarely scavenged. Small pieces of prey thrown on the surface will elicit a better feeding response than a large, whole prey item. After the hatchling's first meal, feedings are only necessary once or twice a week. Constant feeding is not needed and may demand excessive filtration and water changes.

The most important aspect of feeding growing nymphs is maintaining clean water. The

Belostoma flumineum from northeastern Ohio feeding on a pond snail.

The results of one night of egg laying on the back of a male *A. herberti*.

less the young are fed, the less the water needs to be replaced. Adults are less susceptible to dirty water because they store air under the wings where the breathing apparatus is protected from fatty lipids. Although the bugs breathe air from the surface, the lipids in fouled water will eventually coat and disrupt the hydrophobic setae that allow the water bugs to breathe. Food items offered in captivity are seldom small enough to be entirely consumed by the nymphs and leftovers decompose in the water. Even if all leftovers are removed, waste is also normally discharged into the water. The nymphs are aware of poor water quality and are able to shoot waste out of the water, but at that point it is generally too late. The jettisoned waste is seen as dried white or orange splotches on the walls of the cage—the color depends on the chosen prey. Water changes after fouling will not unclog the breathing apparatus.

Small creatures that clean up leftover portions of the meal are commonly known as clean-up crews. These can be very useful in reducing buildup of waste along with regular care, but will not cure overfeeding. Some species of springtails live on the surface of water and will feed on surface scum. Underwater ostracods (seed shrimp) can clean up small quantities of leftovers before they rot. Very small species must be chosen for clean-up crews or the water bugs will eat them. These species are not readily available from vendors but they are common on aquatic plants and can be collected from most freshwater sources.

REPRODUCTION AND DEVELOPMENT

Females of the large giant water bugs, including *Lethocerus* and *Benacus*, glue a mass of eggs to emergent vegetation above the water line (usually from twenty to fifty eggs). The male is the caretaker of the brood and without his attention the ova do not hatch. He cleans, moistens, and defends the egg mass. In close quarters, the male can have a difficult time protecting eggs from the female, especially if there is more than one female. Extra females tear down the egg masses of their competitors.

It is extremely difficult to produce young of this group in laboratory situations, partly because it is not easy to discern males from females. Despite some claims (Dunn 1993), the subgenital plates of both sexes appear identical in the majority of species (an exception being *Kirkaldyia deyrolli*, whose females have a short fringe of setae at the tip). Nevertheless, specimen gender may still be identified in more subtle ways. The wild-collected female's abdomen is often swollen with developing greenish eggs that are visible through the underside of the carapace. Also, wild males are commonly (but not always) a little smaller than females. Size can be very helpful in determining wild-caught specimens, but captive-reared adults usually cannot be sexed by size with any accuracy.

The vicious adults should only be kept together as pairs for mating purposes. If females are not provided a choice of egg laying sites, they drop eggs into the water or hold in the ova and die. Eggs dropped into the water quickly swell to about twice their normal size. By then they are light green in color and dead. Healthy eggs laid on emergent objects are smaller and dark brown with pale opercula. Flat surfaces such as floating Styrofoam, cork bark, emergent rocks and artificial shoreline are not acceptable sites. Upright reeds or dowel rods (1/2″ or 1.25 cm diameter) work, but many females find them unacceptable if not arranged to their liking. It is important to watch out for masses of green eggs at the bottom of the water and to experiment with the placement and proximity of reeds or rods. Females should be removed to prevent egg masses from being devoured while, at the same time, the male must

Belostomatids can remain submerged for fifteen minutes in clean water.

Hydrachna mites often congregate between the legs. Large, darker polyps are approaching the molt to maturity. Small, bright red polyps are growing larvae rather than a different species or genus of water mite.

stay to take care of the eggs. Because of eggs laid or knocked into the water, the difficulties of gestation in artificial habitats severely limit captive rearing of Lethocerinae.

Fortunately the gestation of Belostomatinae ova is not difficult in artificial habitats. Female *Abedus*, *Belostoma*, and other genera glue egg masses onto the thickened outer wings of the male. This behavior makes it easy to provide an acceptable egg-laying site in an artificial habitat. Fifty to a hundred eggs are glued to the males hemelytra within a rubbery mat secreted by the female. The rubbery mat is enlarged with the addition of each egg. The pair spends multiple evenings working together on creating the egg mass. Time required for completion is generally a few days to a week. Often the female only glues five to twenty eggs at a time, then they mate again. More than one female will lay eggs on a male's wings until there is no space left. Competition does not result in cannibalized or drowned eggs unless oviposition is interrupted. Even then only a few eggs would be lost.

The male small giant water bug is likewise caretaker of the eggs. He carries the eggs around and uses his back legs to reach up and massage and clean them. The male also fans the eggs by moving his abdomen up and down through the water's surface. If the water is dirty and low in dissolved oxygen, he holds eggs above the surface most of the time, only dipping them in the water from time to time to prevent desiccation. The *A. herberti* male effectively protects the eggs from the hungry female and most predators. However, if there is very little water and far too many live crickets, a hungry cricket may chew on some of the eggs while their father is looking in the other direction. In the wild, parasitic wasps can insert eggs into them without him noticing. The male is aware of the health of eggs. If ova die because of extremely poor water conditions, he scrapes off the mass and it becomes fair game as food. If the male dies before the eggs hatch, they are very unlikely to hatch. Attempts at saving discarded eggs, or those glued to a dead male, are unlikely to be successful even if moved in and out of the water on a regular rotation. Due to the male's excellent care, large quantities of hatchling Belostomatinae are easily produced in artificial habitats—often far more than one would ever want to rear.

Eggs hatch after about three weeks into wingless, miniature replicas of the adults. The male appears to watch over (or at least not devour) the hatchlings as they emerge, but after a day or so he forgets whether they are his children or his food. In the wild, the tiny hatchlings swim off to hide from predators—including siblings and parents—in nearby vegetation.

Nymphs have some ability to change color to match their surroundings. This is a slow process that takes days, sometimes weeks, though it can be much faster immediately following a molt. All species are able to change from a very pale tan to dark brown. The nymphs of some species, including *Belostoma flumineum* and *Lethocerus americanus*, can develop a green color. Unfortunately this cryptic ability does not interfere with intraspecies cannibalism in captivity because water bugs hunt by movement rather than color. Vertebrate predators like fish and birds may be fooled.

The young molt five times and, if fed often, molt every seven to ten days until the last molt. The final molt requires two to four weeks. Although the abdomen is distended before molting, there are no telltale signs that a molt will occur (with the exception of darkening wing-pads on the subadult before the final molt). While most animals discontinue feeding a few days to a week prior to molting/shedding, giant water bugs eat until the very last minute.

Teneral adult *A. herberti* with the breathing apparatus extended.

Abedus herberti molting to maturity.

After the molt to maturity, giant water bugs live a long time, usually two or three years.

One interesting aspect of rearing belostomatids is the F1 generation often experiences extremely high die-off unless conditions are perfect. However, after this critical stage is passed and enough individuals have been reared to maturity to begin another generation, all subsequent offspring survive significantly better. This is also true for waterscorpions, which are reared in much the same way. This single generation adaptation is not comparable to any other captive invertebrate I have worked with.

HANDLING AND VENOM

Giant water bugs do not feed on nor attack humans but they can bite when stepped on or handled carelessly. Belostomatids are sometimes called "toe biters" but they have terribly inflexible bodies and are unable to bite a toe, unless they are upside-down when stepped on or the head is stepped on in an unlikely way. Giant water bugs are very easy to handle and transfer with little to no risk of envenomation unless one is terribly careless. Even the relatively flexible young may be handled after they are large enough to be held securely by the sides without mortally damaging them. There is little risk of being bitten because it is easy to avoid the short rostrum, though not for lack of effort on their part. If a finger is placed against the rostrum, or if an animal is held tightly in a closed fist, a bite is generally unavoidable. Discomfort caused by a bite from small nymphs is measured in seconds. The bite is immediately painful but, from adults of our large U.S. *Lethocerus* and *Abedus,* usually less so than a wasp sting. The bite of certain tropical gargantuans may be far worse as they are capable of injecting greater quantities of venom.

Notable quantities of venom do not appear to be injected during a defensive bite because there is no observable breakdown of the surrounding tissue on a macroscopic level (pers. obs., *Abedus herberti* and *Lethocerus* envenomations). The venom generally causes a painful

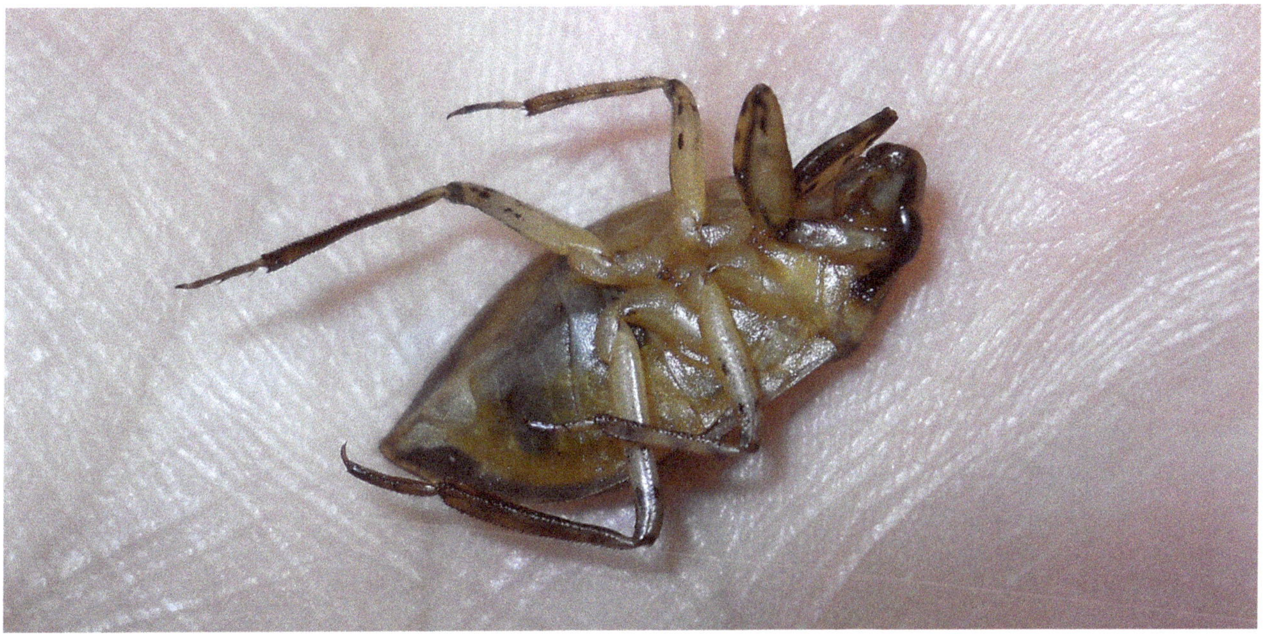

Thanatosis: some species like this *Belostoma* sp. are incredibly accurate at playing dead.

Sponge island in individual nymph enclosure.

bite since it breaks down cells, though there may by pain-causing components unrelated to digestion. The venom is not known for any specific toxicity. Allergic reactions are not currently described in the literature.

Rees and Offord (1969) studied the effects of *Lethocerus cordofanus* venom and state that after 10-15 minutes the interior of the prey is liquefied and sucked out. I do not believe they were aware of the mechanical properties of the belostomatid feeding process. Belostomatids may appear to be passively sucking liquid, as if from a straw, but this "straw" contains a rough-edged whip-like stylet whose mechanical action greatly exceeds the confines of the rostrum. (This can be likened to chewing up one's prey through a hole in its skin using a powered drain snake/drain auger.)

HOUSING

To prevent cannibalism in the close quarters of captivity, it is necessary to place each nymph in its own container soon after hatching. The adult male can be moved to a different cage before the eggs on his back hatch to eliminate initial losses. Otherwise, within a few hours, there will only be one or two surviving nymphs—and only those expert at concealment.

Reasonably sized containers such as 1.5″ (3.8 cm) diameter disposable drinking cups (for the first 3 instars) and 16 oz (475 ml), 3.5″ (8.4 cm) diameter plastic containers (4th instar to adult) have been used to rear thousands of *Abedus herberti*. A tall, 32 oz (950 ml) deli cup, 3.5″ (8.4 cm) diameter, is just big enough to rear each *Lethocerus* nymph while a disposable drinking glass can be used for most individual

Abedus or *Belostoma* nymphs. Even the final molt is successfully completed in a container not much wider than the adult bug's length. If nymphs die in huge numbers, it may be worth providing each a large rearing container such as a 2-gallon (4 L) bucket or aquarium. If placed in larger containers, fouling is rarely a major concern, but the space required to set up hundreds of nymphs, each with a few gallons of water, is prohibitive, while deaths from fouling are not entirely circumvented.

Each rearing container should contain 1.5″ (3.7 cm) to 2.5″ (6 cm) of water and some Java moss or a chunk of foam. I recommend Java moss (*Taxiphyllum barbieri*) as it is a common aquarium plant from Southeast Asia that requires very little light and no care. Foam pieces should be higher than the water level and not take up too much of the floor space. There must be foam or moss available for the nymph to hold on to as it molts. Additionally, there should be some open areas among the sticks or plants, because the nymphs molt into open water while holding onto an object. Lack of open water causes molting deformities. Finally, remember nymphs are far more susceptible to poor water quality closest to the molting period.

The water in tiny cages may need to be replaced one to two days after each feeding because severely fouled water will kill the nymphs. The water can be dumped out and replaced with room temperature tap water. There is no need to remove chlorine or minerals present in tap or well water. Chlorine does, however, kill ostracods employed as clean-up crews. Water changes should be preventative, not reactive, or the damage to the bug nymph will likely have already been done. Furthermore, extremely small nymphs can be mechanically damaged if handled.

Adult *Abedus herberti* and *Belostoma flumineum* are routinely housed together in large

Each of the compound eyes on this adult contain more than 3,000 ommatidia.

groups but may resort to cannibalism if allowed to go hungry for a lengthy time. They can endure a month or two of starvation and only feed on other adults that have starved to death. Adults usually cannot hurt other healthy adult specimens. On the other hand, specimens of a *Belostoma* sp. purchased from Carolina Biological Supply Company in the 1990s rapidly cannibalized (they were sucking the life from each other within minutes of being placed together). Cannibalistic tendencies cannot always be associated with a particular genus. Large *Lethocerus*, including *L. americanus* and *L. medius*, seem rather tolerant of each other initially but cannibalize slowly over time. With adequate food and caging it can take many weeks or months before only a single adult remains.

Even species that do not normally cannibalize as adults will eat specimens with malformed or soft exoskeletons. Although the teneral period visually seems limited to a few hours as the exoskeleton transforms from white to the final brown color, maximum rigidity can take as long as three weeks. The exoskeleton of a newly matured animal must be allowed to harden for at least two weeks before it can be placed with other adults. Specimens with malformed wings may never be safe. The thick, leathery exoskeleton of this stage is the primary defense that prevents cannibalism. No matter how long it has to harden, the shell of the immature stages is relatively thin and soft. Nymphs should never be kept with adults (or each other). No matter how well fed the adults are, they will hunt down and eat the nymphs.

Groups of smaller giant water bug adults are successfully housed in a ten to twenty-gallon (40-80 L) aquarium. No more than eight 1.5-inch (3.7 cm) *A. herberti* or fifteen 3/4" (2 cm) *B. flumineum* should be housed in a partly-filled 10-gallon (40 L) aquarium. Overcrowding should be avoided as it makes it difficult to maintain adequate feeding and water quality. Apparatus for filtration and aeration only becomes necessary when heavily fed or crowded and may still be inadequate. The water depth should be from two to six inches (5-15 cm). Increased depth of water is primarily unusable space and too high a level will make it difficult to contain the animals. Under normal circumstances, the water needs to be replaced just once every month or two when the water becomes yellow or to replace for evaporation. A piece of floating wood, emergent rocks, sponge, or artificial shoreline should be placed in the cage to facilitate grooming and egg care. Artificial shoreline can be made from piling up gravel at one side or constructed with sponge or foam. If the water is crowded with small-leafed aquatic plants, such as Java moss, some hatchlings survive longer till they can be removed. Individual enclosures should be made available to isolate males holding eggs if the hatchlings are to be reared.

Each *Lethocerus* or related species should be kept in its own cage. Half a gallon (1 L) of water per bug is plenty. For mating purposes, a pair can usually be placed together safely for weeks in a half-filled 10-gallon (40 L) aquarium. The mated pair of larger giant water bugs should be provided with emergent water grasses or vertical dowel rods (projecting from the water at least three inches) to provide egg-laying sites. Upright dowel rods can be glued or tightly inserted into holes drilled in a small sandstone or tufa rock. The base should be heavy enough to prevent the structure from being moved about by the bugs, knocked over, or pushed against the side.

Abedus female laying eggs on the male`s hemelytra.

Preventing escape is especially important because water bugs dry up and die within a day or two if they escape from their artificial enclosure. Nymphs cannot fly or climb smooth surfaces, so a lidless cup or 32 oz (950 ml) tall container works very well to keep them contained. Be careful to look for dirt and calcium buildup on the sides, which would allow the bugs to escape. Also, remember that sticks should be kept a few inches below the rim of the container or the bugs simply climb the sticks and walk away. These creatures are acutely aware of human presence and go into hiding when someone enters the room. They also tend to escape at night. The paddle-shaped legs allow for surprisingly skillful climbing of thin cords and tubes. Pre-formed aquarium lids

Belostoma flumineum Say, 1832
This water bug is found in ponds across North America from Canada to Mexico, and supposedly feeds on a wide range of prey. However, the adults of northern specimens seem restricted to snails versus other prey, which begs the question of identification versus geographic behavioral differences. *Belostoma lutarium* and *B. testaceum* are common in its southern range and look incredibly similar. There are seventy or so described species found across the New World and many of these are easily confused because they look similar and are mostly differentiated by dissection of the male genitalia (Ribeiro 2007). Adults are usually less than an inch (~2 cm) long. The husbandry experiences described below are with stock originating from northeastern Ohio.

Nymphs are voracious feeders and accept almost any active prey, but adults appear to be restricted to snails. A variety of prey including daphnia, glassworms, crickets, isopods, and roaches evoked no feeding response from the original wild-caught adults, even after months of starvation. I wondered if they naturally had a very sparse appetite or were trying to overwinter. I ran across a reference to *Limnogeton* and decided to try snails. After a month of providing the correct prey they became more active. I discovered these *Belostoma* were extremely voracious if offered the correct prey and easily ate their own weight in snails daily. Adults began mating and laying eggs at a relatively low temperature, 72° F (22° C), in the middle of the winter (December and January). Not all snails are equal prey. Pond snails (*Physa*) are eaten three times faster than common ramshorn snails (*Planorbarius*). It is difficult to be certain if trumpet snails (*Melanoides tuberculata*) are eaten because of their ability to stay shut for long periods and the opaque shell.

The adults are especially adept at playing dead when touched. Those that are inactive due to low food, or temperatures below 72° F (22° C), commonly will not move for hours after being touched. Large nymphs of *Lethocerus* and *Abedus* are also capable of playing dead, but rarely for more than fifteen seconds.

Like *Abedus herberti*, females will lay eggs year-round but require months of conditioning to return them to a breeding state if they have been underfed or kept below 70° F (21° C) for more than a few days. An egg-laying female can consume two to three small (4-5 mm) pond snails every day for months on end. Males can eat nearly as much. Other prey may be taken occasionally, but will not result in oviposition. One female glues 15-25 eggs per night to the male's hemelytra for up to a week. Nymphs hatch according to when they were laid but usually over a shorter period. They hatch approximately twenty-two days after the eggs are laid at 74° F (23° C). Nymphs are always active and will accept various prey but prefer tiny snails. Full-grown animals can be amazingly sedentary to the point where their living status is uncertain. They are much more active and less likely to play dead for long when fed well. Specimens live two to three years after reaching maturity.

Pieces of sponge can be cut to provide perching areas (*Belostoma* cannot cling to the sides of a plastic or glass container). However, unlike most of the aquatics, *Belostoma* do not climb onto emergent decorations in the enclosure for grooming. The sponge offers a great perch and likely aids in biological filtration due to the surface area, but there is one drawback. It should not be lifted and set back down if the underside cannot be immediately checked for pinned bugs. I have not killed any specimens, but I set a large piece on top of a female for about half an hour before I looked back and

Belostoma flumineum male after first night of egg laying by female (16 eggs).

Egg mass after the second night of laying (29 eggs). Notice the pale eggs were very recently laid. They will darken in half an hour or so. Another two nights and there were 67 eggs, but plenty of room on the hemelytra.

Same egg mass, but the oldest eggs have retained dark operculums and grown significantly in length. *Belostoma* ova expansion is unusual because it is one dimensional (egg grow three times longer, but maintain the same width). Other than belostomatids and nepids, very few true bug have elastic egg shells that expand during development. Egg growth is normal for beetles (Coleoptera).

noticed I could not find her. She looked dead (and not like the usual thanatosis). I set her on top of the sponge where she sat for hours. The next day she was back in the water, but it was nearly a week before she was mating and laying eggs again.

Lethocerus americanus molt to adult. Breathing snorkels at the tail end are extended at this time. Afterwards the breathing apparatus is retracted and hidden beneath the hemelytra.

Electric Light Bug

Benacus griseus (Say, 1832)

This bug is found north to Ohio, southwest to Texas, and southeast to Florida. It is the only member of the genus *Benacus*. This water bug seems to be more common in the southeastern U.S. than *Lethocerus*. Although it looks very similar to its relatives in the genus *Lethocerus*, the adults are highly cannibalistic and often drown in captive habitats. I would not recommend this species for captive husbandry nor displays. Adults may naturally be short-lived compared to other giant water bugs or it may be an artifact of captivity. They may simply not survive long in artificial habitats that support other belostomatids.

Although various giant water bugs are attracted to lights, *Benacus griseus* is the only one I have seen in large quantities. I have seen a few *Lethocerus* at lights and a dozen or so *Belostoma*. On August 20th, 1995, I found more than fifty *B. griseus* specimens over the period of a few hours. The location was around the lights of three gas stations in St. George, South Carolina. Before finding the first one, I went in and asked the clerks if they had seen any big bugs coming to the lights. They said they had never seen any big insects around. I walked out and there was one in front of the door. It is amazing what people do not notice.

American Giant Water Bug

Lethocerus americanus (Leidy, 1847)

Lethocerus americanus, *L. medius* (Guérin-Méneville, 1857), and *L. uhleri* (Montandon 1896) are common in the United States. There are currently 22 described species in this genus including *L. angustipes, L. annulipes, L. bruchi, L. camposi, L. collosicus, L. delpontei, L. dilatus, L. distinctifemur, L. grandis, L. indicus, L. insulanus, L. jimenezasuai, L. maximus, L. mazzai, L. melloleitaoi, L. oculatus, L. patruelis,* and *L. truxali* (Perez-Goodwyn 2006). Keep in mind that *Belostoma* and *Lethocerus* were used interchangeably in original descriptions so many older museum

Lethocerus medius from Arizona.

Lethocerus americanus nymphs can be green or brown in color, subadult from northeastern Ohio.

Dried *Lethocerus* ova on a stick of appropriate diameter, museum specimen.

Tanzanian *Laccotrephes calcar,* captive-bred adult eating a feeder comet. © Alex Yelich

Adult *Lethocerus* can be maintained safely in smaller containers than large nymphs because space is not required for molting. Adult bugs never molt.

specimens of *Lethocerus* will be still labeled *Belostoma*.

Nymphs can range in color from green to brown according to their captive enclosure. In northern Ohio, specimens are found in ponds near water lilies, rarely cattails. They are not found among smaller pondweeds where one finds *Belostoma*. Nymphs of this large species (2″ + or 5-6 cm) grows just as rapidly, if not faster than, the relatively tiny *Abedus* in captive enclosures. The ultimate molt usually takes less than three weeks. Specimens are very hardy and easily maintained. Nymphs are active and energetic when fed, but do not move a whole lot otherwise. All specimens except for a fully mature pair should be kept in individual enclosures. The only difficult part of rearing this species is getting females to lay eggs on an emergent stick rather than dropping eggs in the water to die. As long as they are kept separate, the adult life span is usually two years or more. *Lethocerus medius*, found from Texas to Arizona, is similarly hardy and appears to be identical in husbandry requirements.

WATERSCORPIONS (NEPIDAE)

Members of the Family Nepidae, the waterscorpions, are ambush predators and use their slender raptorial front legs to grab frail prey. They are capable of stalking quarry, but prefer to wait patiently for food to come within reach. Waterscorpions prefer slowly moving or static water with plenty of vegetation to hide and hunt within. They are not adapted for swimming and they cannot walk on water. Their primary adaptation for aquatic environments is their respiratory system with long snorkel that leads to the common name. The waterscorpion

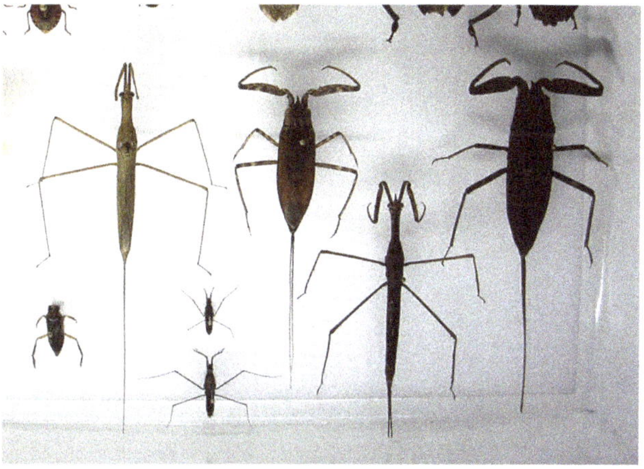

Various shapes and sizes in four waterscorpions.

Ranatra linearis, Ukraine. © Vitalii Hulai

tail is nothing like that of the scorpion—the tail has no stinger, nor venom, and is not composed of muscular abdominal segments.

According to 100-year-old Entomological Society of America common name rules, waterscorpion does not have a space in the middle because that would suggest it is a type of scorpion, which it is not. The rules are simple, just as a butterfly is not a fly and a ladybug is not a bug. A water scorpion would be a scorpion that lives near water, not an aquatic true bug in the Family Nepidae.

Ranatra fusca feeding on a terrestrial isopod or sowbug.

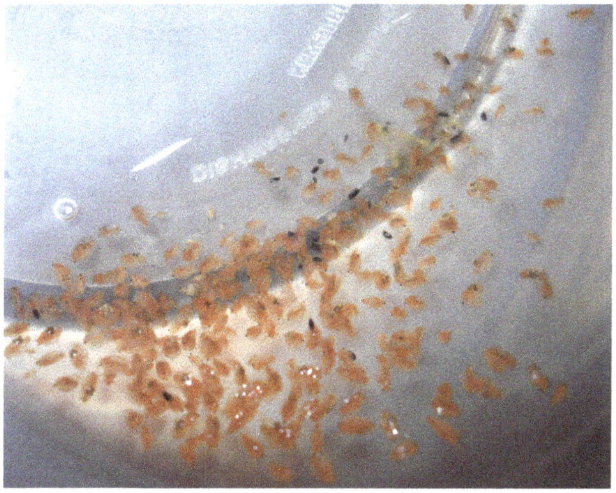

Daphnia magna are a superb cultured prey for early instar waterscorpion nymphs, but will also be consumed by adults.

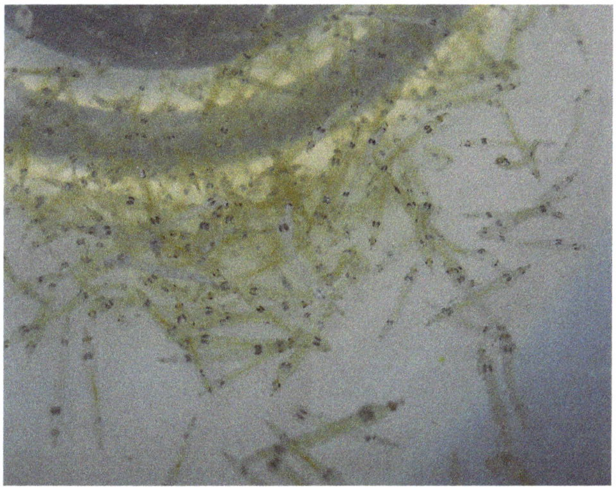

Glassworms, midge larva from the genus *Chaoborus*, are seasonally available as inexpensive feeders from many pet stores. Waterscorpions can usually capture them despite reduced visual hunting clues.

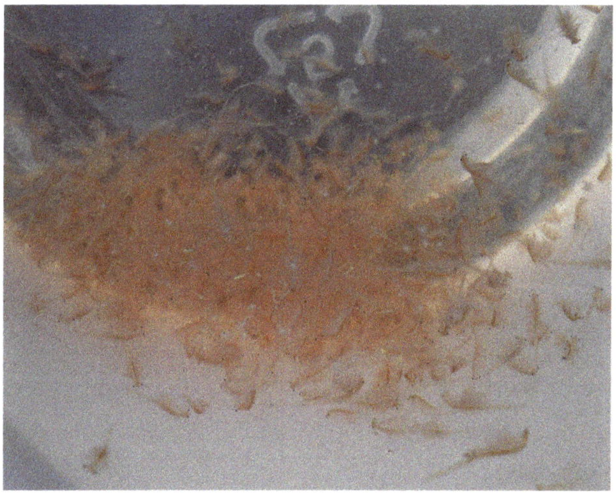

Brine shrimp (*Artemia* sp.) are a useful and inexpensive prey item commonly sold as a feeder. They are difficult and time consuming to rear without large facilities and require brackish to full salt water. Salt water should be rinsed off and they only live a few hours in fresh water. A little salt does not harm the waterscorpions.

Many of the Nepidae resemble a stick or twig (Subfamily Ranatrinae) while those in the Subfamily Nepinae look like an elongated giant water bug or one with a tail. Ranatrinae have a long, narrow pronotum and elongate coxae of the front legs. Ranatrinae are the only commonly encountered subfamily in North America. While there are similar *Ranatra* species found across the globe and into the tropics, Nepinae are more often reported from tropical areas. The Nepinae have a wide, trapezoidal pronotum and short front coxae. Members of this subfamily have a broad, flat body and burly forelegs similar to *Lethocerus*, but also possess the elongated breathing tubes and stilted walking legs of the stick-like *Ranatra*. The United States hosts one *Nepa* species, *Nepa apiculata*, found in the central and eastern states, but it is small and somewhat uncommon. There are half a dozen other *Nepa* species found across the globe. The large *Laccotrephes* from Malaysia and Tanzania have been reared in captivity and have shown up now and then from invertebrate vendors. These waterscorpions have a massive crested pronotum around a tiny head that reminds me of the end of the movie *Beetlejuice* (1988) when the witch doctor shrinks Keaton's head, but not his body. Although they resemble *Nepa,* having a very similar pattern of bumps on the pronotum, they come from a distinctly different genus. The Malaysian waterscorpion appears nearly identical to the Tanzanian, if slightly smaller, but it is a different species as determined by dissection of the male genitalia and minor differences in the miniature, four-segmented antennae concealed beneath the eyes. There are dozens of *Laccotrephes* species found across the (mostly Old World) tropics. They range in body size from half an inch to two inches (1.25 cm to 4.9 cm) (Poisson 1965). Tiny, New World *Curicta* species are more elongate and look like miniature versions of their counterparts in the genus *Laccotrephes*. There are thirteen species found in the U.S. out of 270 worldwide.

Laccotrephes hatching from an egg mass.

Wild-caught adults are often easy to differentiate from those reared in the laboratory. Wild-caught are not only bigger, but they tend to be a little bit beat up and can be missing a small part of the breathing tube, or a tarsus or two. Our native *Ranatra* spp. often have a few, to a dozen or more, parasitic *Hydrachna* mite polyps attached. The larvae are usually fire engine red. The exoskeleton of wild adult *Nepa* and *Laccotrephes* are often embedded with green algae that can make them look especially colorful. Even when kept in well-lit aquariums, captive-bred animals do not seem to pick up the same algae and usually only display the natural color of their own exoskeleton.

DIETARY REQUIREMENTS

The diet of waterscorpions is similar to, but a little less catholic, than that of the ravenous belostomatids. These bugs are not mobile hunters. They stalk quarry if it moves a lot and does not appear too large. They tend to wait until

The freshly molted subadult *Laccotrephes* is very thin, pale, and pink.

This subadult is a few weeks prior to the ultimate molt. Note the small springtails on the water`s surface that help to keep down on surface scum.

anything big has drowned before they consider eating it. Scavenging behavior is limited as feeding is primarily triggered by movement. Small crickets and isopods can be thrown on the surface for feeding. When it comes to aquatic prey, only relatively small and defenseless organisms are heartily accepted foods. Waterfleas (*Daphnia magna*) are an excellent food for nymphs and adults. *Daphnia magna* cultures are readily available and easy to keep but are difficult to rear in the large quantities that would be required to feed large nymph and adult waterscorpions. A large nymph can eat hundreds of full-grown daphnia to get to the next molt (this is why waterscorpions and giant water bugs have been investigated for control of mosquito larvae, which are eaten with similar gusto). Waterscorpions generally do not possess the strength, and lack desire, to pry loose aquatic snails in order to eat them. Small fish and tadpoles are among the common natural prey of waterscorpions but are contraindicated in the laboratory due to excessive leftover material, certain ethical ideals, and lack of necessity (vertebrate prey are entirely unnecessary for full development of nymphs).

1st instar *Laccotrephes* change to brown after half an hour. © Alex Yelich

Offering prey items to nymphs and adults twice a week is sufficient. Keep in mind that overfeeding is usually more hazardous than underfeeding. As with Belostomatids, adults are able to survive some starvation and less clean water than nymphs because their breathing system is shielded by the wings. Nymphs have a different breathing system composed of expanded edges of the abdomen that fold over the abdominal sternites. This is not as protected from fouled water. Waterscorpions usually will not leave the water if conditions deteriorate from overfeeding (while belostomatids climb on emergent surfaces if the water fouls).

REPRODUCTION AND DEVELOPMENT

The females are usually larger and have a more rotund abdomen, but otherwise look the same as males. There is no shape or notable color dimorphism in the sexes. Waterscorpions mate only in the late hours of the evening a few hours after the lights have gone out. Females may be attracted to vibrations from movements the male makes with his raptorial forelegs. The male slowly moves over towards the female, a little behind but beside her. He then nudges the end of his abdomen against hers and if she is receptive they mate. Unlike most true bugs that mate with the partners facing away, Nepidae mate sitting beside each other. (This position seems to be common for predatory groups.) Mating takes about a half an hour. Adults separate quickly if disturbed. Neither parent provides care or protection to the eggs, with exception for placement and the structure of the egg itself.

Ranatrinae deposit their eggs singly into wet porous objects just above the surface of the water (usually in aquatic vegetation), at a rate of two to four eggs each night. Partly submerged compressed fern root, cork bark, or a tight clump of Java moss on a floating platform

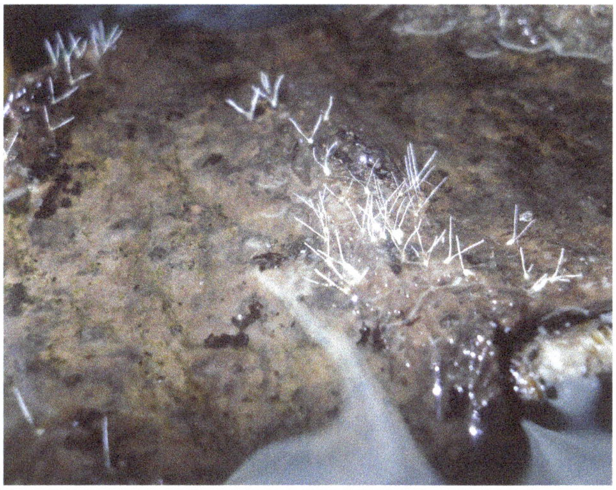

Ranatra eggs laid in cork bark, only the filaments are visible.

Laccotrephes eggs have five filaments on the operculum. © Alex Yelich

can be placed in the artificial habitat for this purpose. Aquatic vegetation such as living cattails may be used in nature, but would be very difficult to duplicate in a small area and without extensive lighting and substrate. Cut pieces may decompose and sink before the eggs have a chance to hatch.

Nepinae egg laying is a little more complex and requires specific substrate depth and consistency on a surface. They oviposit on the shoreline, not in aquatic vegetation. Ova are glued together at the base in a hardened gelatinous material like the egg masses of belostomatids. *Nepa apiculata* form egg masses of one to two dozen, while *Laccotrephes calcar* masses can number from twenty to sixty. Egg mass construction is completed in a single evening. Egg masses will not be added to later if interrupted. In captivity the masses are formed in 1/4" (~6 mm) fine peat on a sponge base, among sticks or moss fibers. The gelatinous material is glued to attachment points beneath the fine peat. Sticks, moss, or sponge surfaces form the anchor. A deep layer of substrate will prevent egg laying since an attachment point is as important as the fine, upper layer of muddy substrate. Without a proper site for oviposition, the eggs will be dropped in the water where they immediately drown. Attempts to rescue eggs from below the water's surface are futile (pers. obs.).

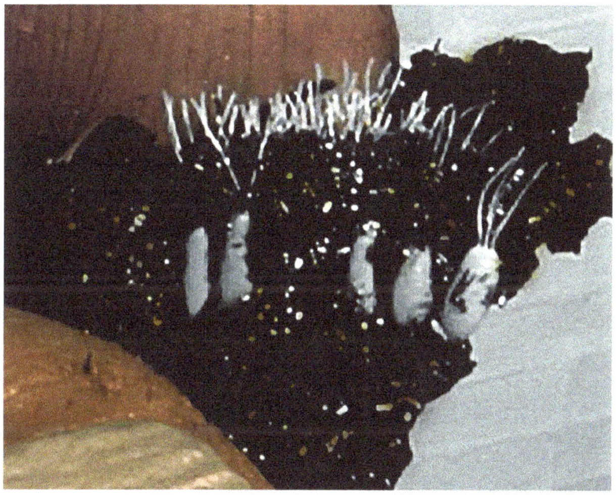

Cross-section of a *Laccotrephes* egg cluster.

Ova would be invisible if it were not for the long, obvious, hair-like filaments projecting out of the substrate. Even so they are still relatively unnoticeable unless one knows what to look for. The filaments, which are part of the structure of the operculum, allow the buried eggs to breathe but do not allow for underwater breathing. It is important the filaments

Adult Tanzanian waterscorpion from captive-bred stock. © Alex Yelich

are not submerged or the eggs will drown. Eggs dropped in the water swell up to twice their normal size and die. Females will eventually drop eggs in the water if they cannot find a good egg-laying site. Even if removed shortly after, before they have swollen considerably, drowned eggs cannot be salvaged. If the eggs are allowed to dry out they quickly desiccate. Operculum filaments tend to be genera specific in number, but can be different lengths according to species (for example in *Ranatra* spp.). The two filaments for *Ranatra* and five filaments on *Laccotrephes* can be seen in the images provided. *Nepa* ova have eleven filaments. *Curicta* species have fifteen filaments on the operculum (Wiley 1922).

Nymphs emerge from the eggs after two to four weeks. The tiny hatchlings look similar to the adults, but they have no wings and possess relatively shorter breathing snorkels. The hatchlings of *Ranatra* are off-white (similar to the color of aquatic roots or dead vegetation) and very thin. They resemble a tiny clump of roots more than a stick. *Laccotrephes* emerge fire engine red in color (*Nepa* and *Curicta* yellow) but change to brown after a few hours. Unless kept in a relatively large and well-planted cage, hatchlings should be kept separate from adults so they are not eaten. Hatchlings are cannibalistic, but can be kept together since they do not aggressively chase after and attack each other. If kept together it is important to feed the nymphs every other day and offer space and perches. Otherwise, they should be kept separately.

HOUSING

Two-and-a-half- to twenty-gallon aquariums make excellent communal enclosures depending on the number of adults and the species (aquarium dimensions L x W x H: Two-and-a-half-gallon = 12″ x 6″ x 8″ (30.5 cm x 15 cm x 20 cm), Five-gallon = 16″ x 8″ x 10″ (40.5 cm x 20 cm x 25 cm), Ten-gallon = 20″ x 10″ x 12″ (50 cm x 25 cm x 30.5 cm), Twenty-gallon long = 30″ x 12″ x 12″ (76 cm x 30.5 cm x 30.5 cm)). Adults are normally kept together and do not appear to ever feed on each other unless one dies. Although they have rather small wings, nepids can fly. I have never seen a waterscorpion fly. (There is such an insect as a scorpionfly but it is not a type of fly nor waterscorpion.) I have encountered a specimen that must have landed in an aboveground pool and another in a water slide. Nevertheless, a lid does not appear to be necessary because they cannot take off very well or lack specific environmental cues (such as moonlight and temperature cycles). An easily scalable surface like floating wood or emergent plants should be provided to prevent drowning. Because *Ranatra* and *Nepa* are very susceptible to poor water quality, a 30% water change should be practiced every two to three weeks. An aquarium air stone (diffuser) can be used to increase oxygen in the water. The air pump should be adjusted to a low setting to provide a gentle flow. Waterscorpions can climb airline tubing.

A ten-gallon aquarium can be set up with just under 2.5″ (6.4 cm) of water, some Java moss, and two 2.5″ (6.4 cm) x 10″ (25 cm) pieces of 2.5″ thick sponge. A little Java moss is placed on the sponge and then covered with half an inch (1.3 cm) of peat from peat pellets. Pellet peat is usually much more finely ground than standard peat and offers a muddy consistency without muddying the water. The tank is filled with water to just below the top of the sponge and the sponges are placed lengthwise at each end. Java moss aids the bugs in climbing to prevent drowning. The peat layer provides a media for egg laying and the sponge keeps the peat wet while above the water level. No air stones are necessary. The sponge should not be moved because it disturbs the peat and moss growth. More importantly, large pieces of sponge accidentally set on top of an animal will pin it and lead to drowning.

Waterscorpions usually play dead when handled. The behavior is known as thanatosis and is a common defense across various insect orders and other animals.

HANDLING

Waterscorpions feign death when picked up and rarely move around when held. Even when active they are not agile and do not attack human fingers. The only way to be bitten is to place the rostrum against the skin and wait. While giant water bug bites are unlikely in the face of minimal efforts at prevention, accidental waterscorpion bites are nearly impossible. The rare bite reports seem highly dubious since it would take a good bit of effort on the victim's part, but a bite would likely feel similar to a giant water bug's bite.

SELECT NEPIDS

Laccotrephes calcar Distant, 1904

There are a few dozen very similar appearing species in the genus *Laccotrephes* from Africa, tropical Asia, and Australia. This species is found in Kenya and Tanzania. If measured 'claw to tail' like true scorpions, a large adult would be nearly five inches (~12 cm) long though the body length is less than 2″ (< 5 cm). The underside of the abdomen ranges between dull red and blood red in wild specimens, but is rosy to light tan for captive-reared. Although waterscorpions are generally unknown for defensive secretions, these will produce a milky white fluid from glands in the pronotum if sufficiently disturbed. The fluid has a strong chemical smell but it is nether pungent, nor offensive, and likely the defensive mechanism is bad taste.

Laccotrephes eggs have five straw-like projections sticking out of the top. Ova are buried and these filaments allow them to breathe. Though egg masses can be over an inch across and 1/4″ high (2.5 cm x 0.6 cm) they are not immediately obvious. Masses look like a pad of tiny hairs sticking from the dirt and are only visible if they are sticking out of dark colored substrate.

Egg masses (usually from fifteen to thirty eggs, but up to sixty) are laid in the peat. Unlike *Ranatra*, the egg masses are glued together at the base in a gelatinous material like giant water bug eggs. Egg masses are placed just above the water line on a sloping gravel bank at one end. They hatch in thirty days at 72° F (22° C). Hatchling color can be a surprise for those familiar with other Nepoidea. Fiery red hatchlings pop out of the top and make their way to the water. Within a few hours they transform to chocolate brown.

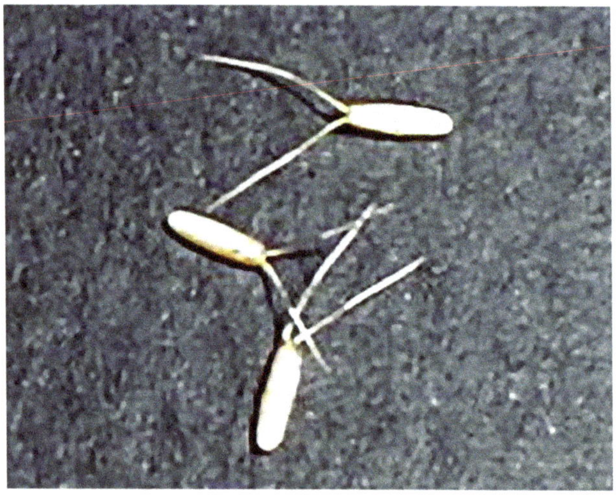

Ranatra quadridentata ova removed from egg laying material, each with two filaments as standard for this genus.

Though adults are safely kept in large groups, nymphs must be removed and reared separately. *Laccotrephes* nymphs do not viciously hunt each other down like giant water bug nymphs, but they are more active than *Ranatra* and will kill each other as they become hungry and wander in search of prey.

Mature specimens will eventually eat every nymph left in the cage. Nymphs are moved to separate cups for rearing. They change color to match the bottom of the container. In white containers, they become translucent following the first molt and do not turn brown again until the molt to adulthood. A bit of Java moss and 1/2" (1.3 cm) of water is placed in each container. Too much Java moss or too little water and they cannot molt correctly. Too much water or too little Java moss and they drown. Daphnia can help keep the water clean though small waterscorpions will eat all the daphnia.

Laccotrephes are aquatic, and can swim, but they live where water and land meet. Drowning can be a major problem in captivity. Inability to reach the surface and dirty water both may result in drowning. They drown if the container is filled too high. Hatchlings are the most vulnerable because they have a relatively short breathing tube. (This snorkel may resemble a straw but is formed by two appendages that can be moved independently.) The required water depth in the container changes with each molt because the breathing snorkel starts out short but grows disproportionately longer with each molt. 100% water changes using aged tap water is necessary if the water looks syrupy. Nymphs store air under expanded flaps folded beneath the abdomen. Adults store air under the wings. Below-body air storage is why nymphs easily suffocate with poor water quality. Adults can survive a lot of abuse. Young grow rapidly if fed daily, but water quality suffers. Feeding once a week is safer but never less often. A large container like a 2-gallon (4 L) plastic shoebox for each nymph might bring nymph die-off close to zero since a larger volume of water is easier to keep clean.

Rearing containers used for older nymphs should be at least four inches (> 10cm) in height. I used tall, 32 oz (950 ml) deli cups, 3.5" (8.4 cm) diameter, to rear hundreds of adults over more than a dozen years. *Laccotrephes* cannot climb the smooth sides of shorter plastic containers, but every once in a while they seem to figure out how to stand on their "tail." They jump containers and kill other nymphs or end up on the floor and dry out. Like crabs, escapees can survive many hours or days, but the breathing apparatus is mortally damaged. When found and put back in

Pair of *Ranatra fusca* from Girard, OH. Note the *Hydrachna* sp. mite polyp near the base of the pronotum of the smaller male (front).

Ranatra australis from Orlando, Florida.

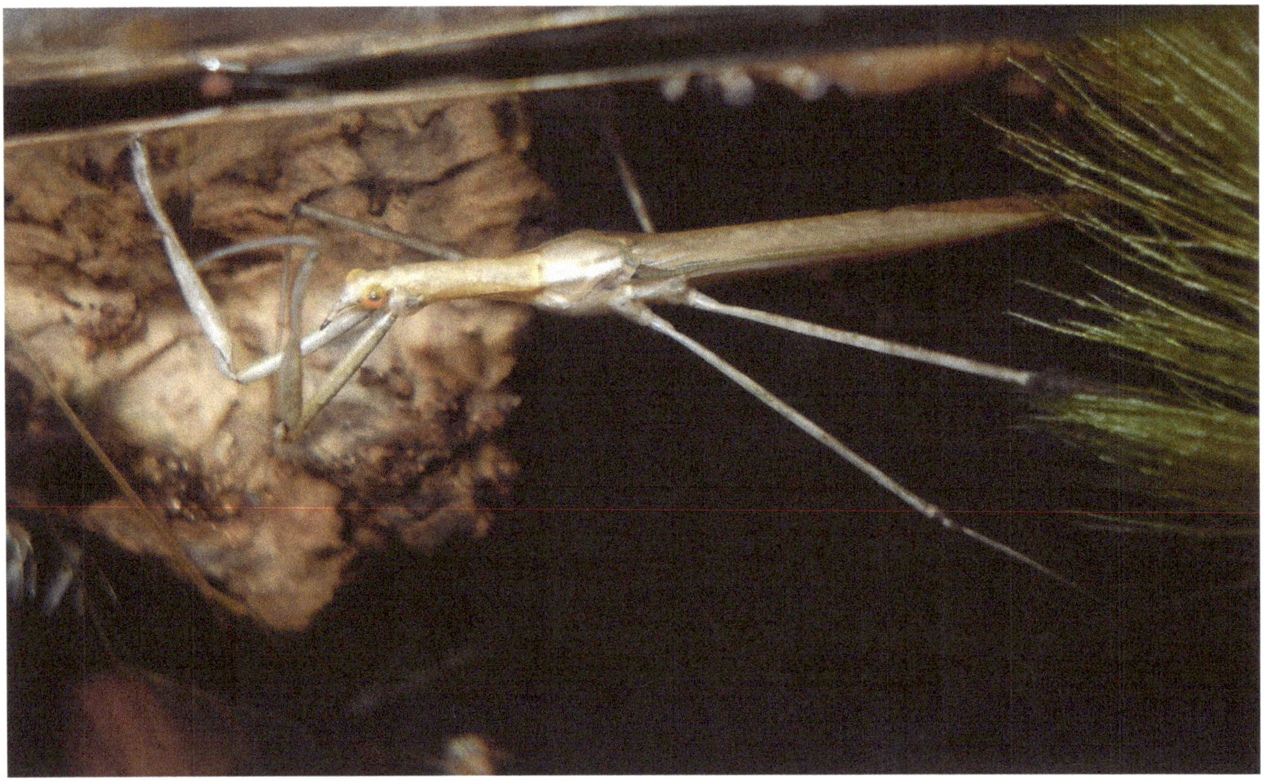

Adult *R. quadridentata* from southern Arizona.

Face of a large *Laccotrephes* nymph.

water they may look okay, but will be dead by the next morning.

Hatchlings can be fed prey larger than themselves. *Laccotrephes* are more aggressive than other waterscorpions. Hatchling nymphs can be started with fruit flies, terrestrial isopods, mealworms, daphnia, crickets, cockroaches, meal moth caterpillars, etc. They seem meek, but throw in small, energetic prey and they come to life. They will feed on gigantic prey, but the leftover carcass should be removed. Large nymphs and adults capture tadpoles and small fish, though this is entirely unnecessary in captivity and resultant fouling can be detrimental to survival. The thirteen-year stock pictured was never reared on vertebrate prey.

Nymphs molt a total of five times after hatching and reach adulthood in ninety days under perfect conditions. The first molts take as little as a week though the final molt takes a minimum of one month. Prior to the last molt, wingpads turn tan, then black. Three to four days after wingpads darken, the molt to adulthood follows.

In captivity, Tanzanian waterscorpions lay eggs year-round and incredible numbers can be reared in a short period. However, it can take time to work out all the "bugs." Females push peat into the water so it has to be replaced regularly. With minimal feeding, months can pass without eggs. At times the females drop all the eggs in the water. (Eggs are dropped if the water level gets low, the wrong type or not enough substrate is used, or there are too many plants or decorations in the water.) Eggs do not develop correctly when the water level is much below them, even though they do not dry out. Cannibalism takes its toll because hatching is either feast or famine. Nymphs reared together in small containers from resultant space limitations can lead to high losses. Still, it can be easier to rear 1st instars together than to attempt individual containers resulting in overextension and neglect. It is easier to run out of flightless fruit flies when fed to large numbers of individual containers because the flies run off.

A wild-caught female waterscorpion from Tanzania that lost a leg and did not survive shipping, 2002. All other photos in this text represent varying generations resulting from the original group of ten adults acquired from this import. Consecutive cultured generations through 2014.

Wild-caught adults are generally easy to sex since females are larger. Nevertheless, gender in captive-reared specimens cannot be determined using size. The underside of a mature female's abdomen becomes rounded within a few weeks, while the male's remains keeled. Captive-bred adults live two to four years. Despite greater longevity, once they exceed two years of age there is little chance of further reproduction.

Laccotrephes do not attack fingers placed in the water and hold their bodies stiff when picked up. They are easily handled without trouble, but they seldom play dead for more than thirty seconds. This family is said to lack the stink glands common to the majority of true bugs, but *Laccotrephes calcar* produces a white defensive fluid with a chemical odor similar to the defensive fluid of *Lethocerus*. When secreted, it usually coats half or more of the pronotum. A number of the predatory, aquatic true bugs can inflict painful bites when handled carelessly. However, the only way *Laccotrephes* can bite is to purposefully force the rostrum into flesh.

Ranatra fusca Beauvois, 1820

Various species across the U.S. look incredibly similar and most can only be properly identified by dissection of the male genitalia. *Ranatra fusca* is most common across the northern states and Canada. *Ranatra quadridentata* is common in Arizona, *R. brevicollis* is found in California, while *R. australis* is the most common species in Florida.

The stick-like *Ranatra* are commonly kept and can be easy to find in ponds and lakes. Specimens used to commonly arrive at pets stores with feeder guppies grown in Florida fish farms.

Nymphs are fragile and rarely move around the cage, which limits cannibalism if they are provided room and adequate prey. (*Daphnia* is highly recommended for early instars more than anything else.) Nymphs molt about once a week, though the ultimate molt takes two weeks; it is possible to get an animal from hatching to adult in fifty days. Immatures are very fragile and prone to dying from water that does not appear old. Weekly 30% water changes are suggested.

The captive habitat requires no emergent areas for egg laying. *Ranatra* are not nearly so picky as the shoreline egg-layers in the Nepinae and deposit eggs on nearly any floating, semi-porous object. Adults live more than a year, seldom two. Feeding ceases below 65° F (18° C).

Ranatra sp. feeding on an aquatic isopod. Growths near the base of the legs may be aquatic mite polyps. (cc) Thomas Bresson

OTHER WATER BUGS

Although aquatic, the vast majority of bugs living in or on the water can drown. They must breathe air. This is an important consideration relative to housing and transportation. When it comes to housing it is important to remember that surface scum, which upsets the water's surface properties, can disturb the balance between the hydrophobic and hydrophilic setae, without which they cannot breathe nor float. In regards to shipping, this means all bugs should be transported dry (on damp paper towel or aquatic plant parts). If placed inside a cup or bag with water and no emergent areas, the constantly moving water can lead to drowning.

Despite the very different appearance of the four families discussed here and the previous chapter, there are two appreciable similarities:

1. The body is adapted for shoreline environments; a semi-aquatic rearing cage is used (clean air and water are likewise required).

2. All species are predatory, but will scavenge on dead insects and animal protein.

Even though numerous true bug families inhabit water environments, only two others are commonly kept by enthusiasts. They are the water striders (Family Gerridae) and the backswimmers (Family Notonectidae). Although shorter and smaller than the scrawnier members of the preceding families, large members of these two families approach an inch in length and consequently are massive enough to provoke the interest of the general hobbyist. Furthermore, water striders and backswimmers are energetic, common, and easily collected in most areas. Adults are usually kept in colonies.

Keeping notonectids and gerrids in the same cage has proven successful for display. It is possible to set up a split paludarium with colonies of these and the shore-loving toad bugs. On the other hand, waterscorpions and the more vicious giant water bugs massacre these smaller water bugs and each other.

WATER STRIDERS (GERRIDAE)

Water strider bugs inhabit the surface of nearly every body of water from the open ocean to tiny placid freshwater puddles. Members of this family are difficult to confuse with other insects. The only other insects normally found racing across the water's surface are whirligig beetles. The oval and seemingly legless whirligigs are obviously unrelated. The water strider's body is thin and usually stick-like. The rear and middle legs are longer than the body and held out to give the bug an overall appearance of an X. The front pair of legs are short,

Aquarius walking on the water, Hinckley, Medina County, Ohio.

In larger animals it is easy to see the water tension holding the animal up.

Water strider, *Gerris* sp. Montreal, Canada. © Martin Gamache

Water striders have notable antennae and large eyes. © Henry Kohler

and raptorial like a praying mantis or waterscorpion. These legs are partially folded and so are not very noticeable. Only the outspread feet touch the surface of the water and these are covered with hydrophobic (water repellant) hairs to easily support these bugs on the thin surface film. They have small, retractable claws on their feet used like brakes to stop forward momentum (Morgan and Thoms 1994). Most Gerrid species are capable of flight. Some species are apterous while in others only the adult male possesses wings. The total number of water strider species comprising the family Gerridae is around 750—close to fifty of these live in the United States. The family Veliidae (smaller water striders) has a thousand member species worldwide.

DIETARY REQUIREMENTS

Water striders feed on live and dead insects that fall onto the water's surface. They have the ability to jump a few inches or more into the air but use this ability primarily as a defense—less commonly as a method of prey acquisition. They do not dive underwater to catch food. Any food that sinks is lost. Live moths, freeze-dried mealworms/crickets and just about any buoyant insects are excellent foods. Size of the prey is unimportant and large numbers of water striders in various stages share large prey items. Floating animal-based fish foods are eaten, but tend to sink too quickly. I have reared two small species of water strider from northern Ohio on fish flakes supplemented by the occasional drugstore beetle, *Stegobium panaceum*, or meal moth.

Healthy animals may feed on a sickly individual. Despite being fed on, the sickly animal might recover if removed. Gerrids are not known to bite people.

REPRODUCTION AND DEVELOPMENT

Water striders glue eggs to floating aquatic vegetation, sticks, or rocks just below the surface of the water. The minute, off-white eggs are thin, elongated cylinders barely exceeding 1 mm in length. With small species the ova can be very small and difficult to locate. Eggs hatch after a few weeks and the tiny hatchlings successfully avoid the hungry maws of the adults for a time. Different stages can be kept together safely if they are similar in size. They are gregarious by instar and species. As with other bugs, five molts are required to reach adulthood. The nymphs only spend about two months (Morgan 1996, pers. obs.) in reaching adulthood. The early instars can molt in as

Gerridae enclosure.

Gerridae enclosure, primarily adults.

Water strider adults and nymphs.

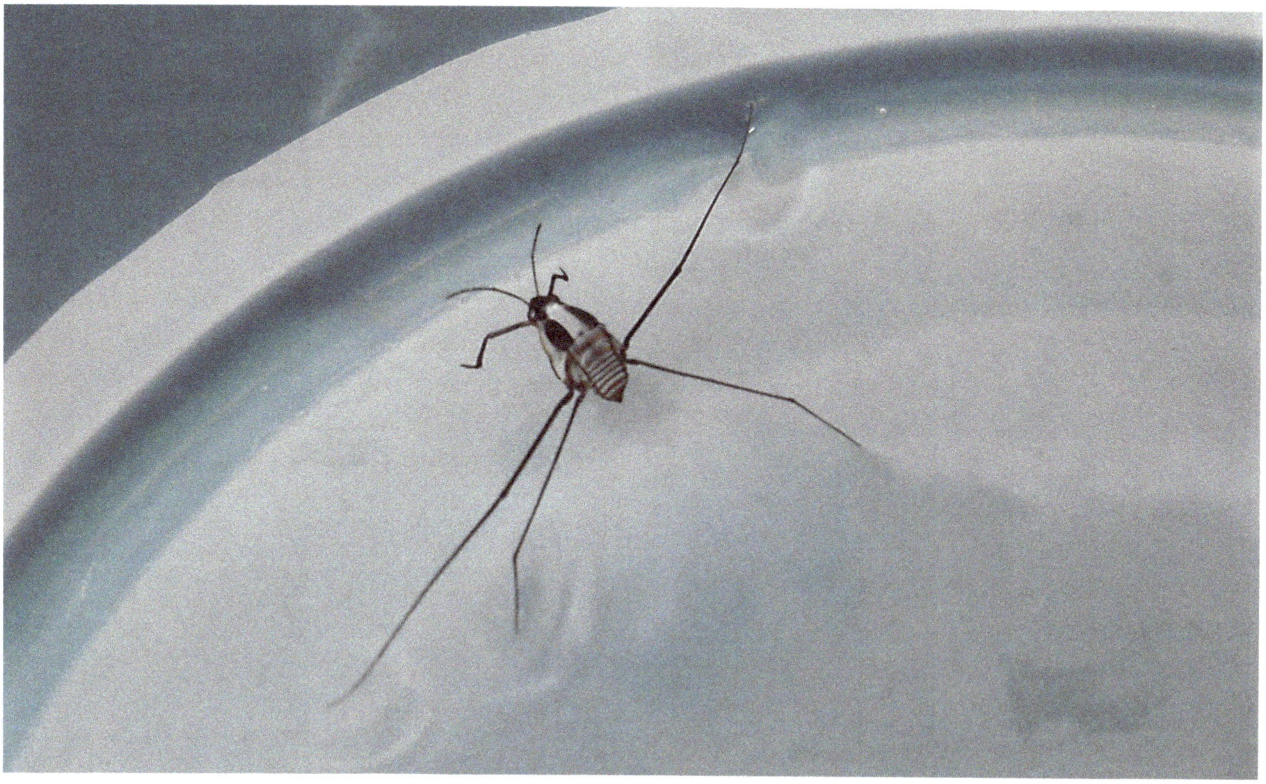

Water strider *Metrobates hesperius* relaxed pose, Hinckley, Ohio.

Water strider creating larger ripples as it pushes against the water`s surface.

Mating pair of water sriders, Bethlehem, Pennsylvania. © Henry Kohler

little as five days with the last molt taking about two weeks. Nymphs are capable of molting on the surface of the water, but bad molts are likely. Emergent decorations are required.

In temperate climates, large nymphs and adults hibernate under wood and leaves near the water's edge. Long before the first snow and even before most leaves have fallen, they search out hibernating spots and disappear from sight. Other types of adult bugs that hibernate are still found crawling around late in the season. Though they can be incredibly common on the surface of ponds and creeks in summertime, Gerrids are impossible to find even in early fall. In the lab, hibernation is unnecessary. They can be kept at room temperature (72-75° F or 22-24° C) year-round. Adults can live more than a year.

HOUSING

Water striders require decent water quality and a secure cage. A ten- or twenty-gallon aquarium makes a good cage. They are kept in groups, but not of different sizes and never of different species. The enclosure should include a sponge filter or undergravel filter to provide filtration. Filters can be powered efficiently by a small, inexpensive air pump purchased from a large department store or pet shop. Using a powerhead with either type of filter is certainly overkill and does not break up the surface scum unless the air venturi feature is utilized. Despite living only on the surface, water striders require fairly clean water because the grimy surface film of dirty water can suffocate or drown them. Captive Gerridae prefer the surface of flowing water, without an extremely turbid surface, but can do fine on clean, static water. The cage should include emergent decorations for Gerrids to perform grooming activities.

Water strider (*Gerrinae* sp.). (cc) Donald Jusa

The cage must also have a well-fitted lid. While they seem to be poor climbers, nevertheless water striders can jump against the sides and climb using water tension. This looks very clumsy compared to how other invertebrates scale glass, but it is effective. Water striders are expert jumpers, often from 2″ to 6″ (5-15 cm) vertically. Many species can fly.

BACKSWIMMERS (NOTONECTIDAE)

These are relatively large insects, not quite an inch (< 2.5 cm) that are probably most familiar as they end up in swimming pools on a regular basis and possess a rather memorable bite. The chlorine does not bother them but eventually they will leave or starve. The backswimmer

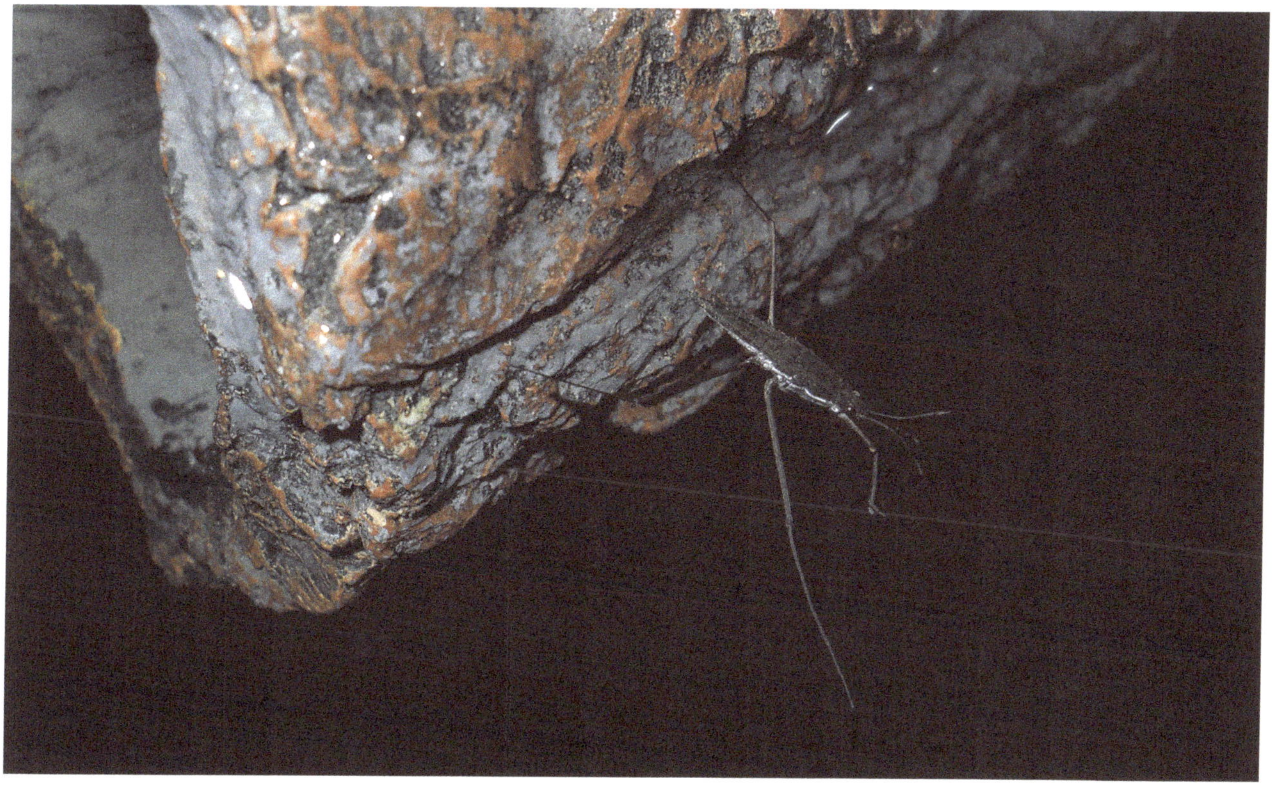

Floating decorations in the enclosure are important for resting, molting, and egg laying.

Notonecta sp. hemelytra coloration, Hinckley, Ohio.

No antennae and very long back legs, *Notonecta* sp. Medina County, Ohio.

Whirligig beetles cannot be mistaken for water striders.

body is boat-shaped while the middle and front legs are about the same length and much shorter than the rear legs. The back legs are as long as the body and flattened into oars. There are around 620 species found in thirty-five genera across the globe; thirty-five of these are residents of the U.S. Backswimmers spend much of their time swimming underwater but

Notice the aqua lung created by setae on this *Notonecta*`s hemelytra. The thin air bubble give it an amorphous silvery appearance.

also spend a bit of time resting just below the surface while exchanging air or clinging to underwater objects. They are superb and rapid swimmers and can stay submerged for hours at a time. Unlike most other water bugs, they have no ability to climb glass or plastic.

Another family of true bugs closely resembles the backswimmers (Family Notonectidae). Water boatmen (Family Corixidae) not only appear similar but are closely related. The two families are comparable in many ways but are easily differentiated. Backswimmers, as the name suggests, spend their time under water upside-down while water boatmen swim and rest upright. Another easy identification clue is the shortened, hairy front legs of the Corixidae, which are used to strain microscopic particles from the water. (They are mostly filter feeders or feed on plants, whereas all the other aquatic hemipterans are predatory.) Lastly, backswimmers average a bit larger than water boatmen and their eyes are smaller in relation to the head. Water boatmen are fascinating creatures that may be easier to propagate due to their more catholic diet and may be more interesting since males can chirp. (They rub their penis against the abdomen to make noises to attract females and the noise is loud enough to be heard even above the water.) Commonly seen species are rather small and attract little attention for research or displays. Older texts referred to notonectids as water boatmen and backswimmers, and designated corixids as "lesser water boatmen" (Stanek 1969, McGavin 1993).

DIETARY REQUIREMENTS

Backswimmers are primarily predatory, but are also successful scavengers. They feed on dead and live insects, small aquatic animals, and most fish foods. They can be kept with water striders to "clean up" sunken food. They are not big eaters. Anyone who has grabbed one of these bugs is familiar with their bite. The bite can be more disagreeable than a bee sting, but the pain lasts less than a minute.

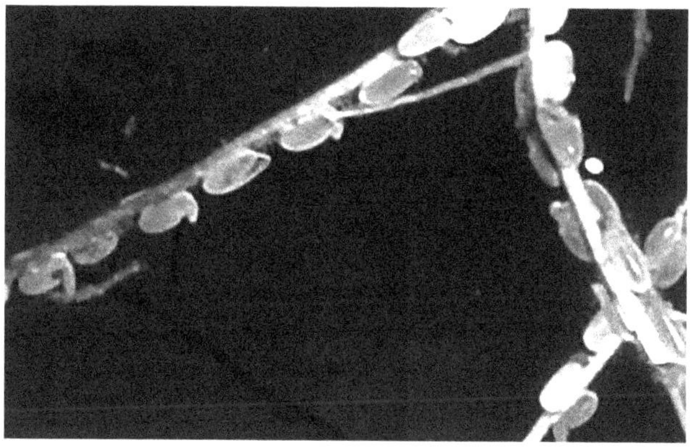

The "bubbles" on these plant roots are actually backswimmer eggs.

Young backswimmers must be reared separately because the adults as well as other nymphs eat them. On the other hand, adults live together peacefully.

REPRODUCTION AND DEVELOPMENT

Females attach eggs to submergent vegetation and other underwater objects. Eggs are generally very easy to see. They hatch after a few weeks and the tiny hatchlings resemble adult *Daphnia*. Nymphs grow quickly through five molts to adulthood. Adults live one to two years in captivity. Most species hibernate through the winter (in temperate climates) in the adult stage.

HOUSING

Backswimmers require good water quality as far as adequate oxygen and limited organics are concerned (a little free chlorine or salt does not bother them). Small enclosures can be used for long periods. Large enclosures should include a sponge filter or undergravel filter to provide filtration. Filters can be powered efficiently by a small, inexpensive air pump purchased from a large department store or pet shop. The cage

Mating *G. oculatus*, Lorain County, Ohio.

Gelastocoris oculatus feeding a cricket.

should include aquatic plants for the backswimmers to lay eggs on and cling to. A secure artificial habitat with a well-fitted lid is suggested. They are poor climbers but are capable of flight. Emergent areas are necessary only if water current is part of the habitat design.

TOAD BUGS (GELASTOCORIDAE)

These medium-sized, semi-aquatic bugs are found along the shoreline of ponds and only rarely streams or rivers. There are seven species in the U.S. and approximately 110 worldwide. They are encountered among small rocks and gravel rather than mud or sand shorelines. Toad bugs jump into the water when frightened by predators, but do their hunting on land. Adults overwinter in nature beneath rocks and can live a year. In captivity, at room temperature, mature specimens seldom live longer than nine months. Nymphs and adults feed on any active prey including springtails, crickets, and isopods. Adults can be housed together safely but nymphs must be reared in individual containers. Nymphs greedily feed on their brethren. Each can be reared on wet sand or gravel in a 5 oz (150 ml) plastic cup with a dozen or more thumbtack holes for ventilation. They do not require pools of water.

Gelastocoris oculatus oculatus (Fabricius, 1798) is a common North American representative of this group. The life cycle was studied by Brown (1994) who found the incubation period averaged nine days, while the five nymphal stadia averaged ten, eight, ten, seventeen, and twenty days respectively at ~84° F (29° C). In my experience hatching takes twelve days and growth is similar, except the fourth instar takes about four days less and fifth instar four days more, at room temperature (72° F or 22° C).

Notonecta glauca eating a bloodworm. (cc) Tom Phillips

Toad bug (*Nerthra* sp.), Western Australia. (cc) John Tann

Milkweed bug (*Oncopeltus fasciatus*) nymphs. (cc) Katja Schulz

SEED BUGS (LYGAEOIDEA)

Seed bug identification can be a bit convoluted and most were considered members of the Family Lygaeidae prior to the turn of the millennium (Henry 1997). Authors may split the seed bugs into three to ten families (and may also award each family its own common name) (McGavin 1993, Eaton and Kaufman 2007). The three main families are usually Lygaeidae, Pyrrhocoridae, and Rhopalidae, but can include the Alydidae and Larginae. The Pyrrhocoridae are differentiated from Lygaeidae because they lack ocelli and the scent glands are often reduced or absent. The Rhopalidae are similar to the Pyrrhocoridae but have slightly different wing venation. If one recognizes the Rhyparochrominae as a Lygaeidae subfamily, visual identification can become even fuzzier. A couple of common species in this group (specifically *Myodocha* spp.) possess reduviid-like front legs, long narrow necks, and violin-shaped bodies. Alydidae species have well-developed scent glands. However their bodies are slender and the heads are often as wide as the pronotum. Larginae have different wing venation from the above families and absent ocelli; some resemble ants and have broad heads like the Alydidae. Alydidae and Larginae are sometimes included in the Coreidae (leaf-footed bugs, next chapter).

In this chapter, we will skip over the finer points of taxonomic hierarchy, itself in a constant state of flux, and consider them all seed bugs like they used to be (Lygaeidae or Lygaeoidea). Commonly encountered species can be cultured under similar conditions. The following attributes can be used to characterize the "seed bugs." All species have four segmented antennae and a simple four-segmented rostrum. The rostrum is long, one-half to two-thirds the length of the body. It extends at least to the end of the thorax and usually past the first abdominal segment. Wings nearly always cover the sides of abdomen (except for short-winged or apterous species). Space between the hemelytra veins is double or more the width of the veins. Overall body design is similar; uniform legs, rostrum long and thin, elongated oval shaped body, and a scutellum covering only about ten percent of the abdomen. Although some of the tiny plant bugs (family Miridae) resemble seed bugs at a glance, plant bugs are only about an eighth of an inch long (0.3 cm) and seed bugs tend to be at least a third of an inch long (0.8 cm).

While this group worldwide contains from three to four thousand separate species (depending on what one considers to be a seed bug), only a tiny number of species are kept by the hobbyist. Part of the reason is certainly due to their small size; the largest seed bugs are only around an inch long (< 2.5 cm). Although Seed Bugs are not very big, they are brightly colored and gregarious. An individual is seldom

A group of beautiful *Oncopeltus fasciatus* drinking from a fruit jelly cup.

Nymphs, *O. fasciatus*.

Mating pair of adult *O. fasciatus*.

Small milkweed bug *Lygaeus kalmii* adult.

impressive, but large groups are eye catching. Seed bugs are harmless display animals that cannot bite, sting, or spine people. The large milkweed bug *Oncopeltus fasciatus* is the most familiar member of this group and has been commonly studied in biology classrooms and reared for insect displays for decades. It is the featured insect on the cover of a popular insect guide (Eaton and Kaufman 2007).

Nearly any keeper can collect at least a few colorful species from nearby fields, gardens, or patches of weeds. Large clusters of specimens can be especially common on wild milkweed (*Asclepias syriaca*) or golden rain tree (*Koelreuteria paniculata*), a landscape plant. Over three hundred seed bug species are found in the United States. Some species are plant pests and can be easily collected because they are nearly cosmopolitan. A notable aspect of seed bugs is the existence of comparable species found throughout the world. There are different orange-and-black species of *Oncopeltus* found on every continent where milkweed grows (McGavin 1993).

DIETARY REQUIREMENTS

As the name suggests, seed bugs feed primarily on seeds and seed pods though all species are somewhat omnivorous. Both of the common North American milkweed bugs, while adapted to feed specifically on milkweed, can be adapted to feed on hulled, raw sunflower seeds in captivity. They cannot get through the tough outer hull of the seed, so it must be removed. Boxelder bugs (*Leptocoris trivittatus*) and the closely related red-shouldered (or) golden rain tree bug (*Jadera haematoloma*) also eat sunflower seeds, but should further be provided hulled pumpkinseeds in their diet.

Small milkweed bug culture container, *L. kalmii.*

Small milkweed bugs are not terribly picky on their host food, *L. kalmii.*

This is one of the predatory, or at least carnivorous pyrrhocorids, that may have potential as display specimens or control of pests, Panay Island, Philippines. © Christian Schwarz

Most seed bugs grow well on these two readily available seeds. Fruit jelly cups (commonly seen as insect or beetle jellies) provide limited nutrition along with needed moisture. Seed bugs will also feed on fresh snap peas and snow peas, but these foods are more expensive, perishable, and may not provide adequate nutrition. Wild-collected stock can take a few generations (frequently with notable die-off) to become accustomed to these alien foods. It is usually not possible to switch back and forth between alien and natural foods without die-off.

Seeds can be kept in a dish so they are easily removed when they start to shrink or mold. Food items should be replaced when they appear shrunken and malformed, usually after two to three weeks. Even afterwards, the seeds only look a little smaller than when they were fresh. It is surprising how many seed bugs can be reared to adults on just a few small seeds. Many types of seeds, including sunflower and pumpkin, can be purchased cheaply in the bulk sections of grocery stores. Less than a dollar in sunflower seeds will sustain hundreds of large milkweed bugs from egg to adult. Roasted seeds are easier to find at most stores. Roasted sunflower, pumpkin, and peanuts are fed upon,

Jadera haematoloma drinking from a carrot.

"This second predatory *Dindymus* species is more common. They usually live on the forest floor, but occasionally are seen higher on the vegetation. Pairs are often seen running around tail to tail in copula, with the larger female dragging the male. Their tandem warning coloration is eye-catching." Panay Island, Philippines. Pers. comm. and © Christian Schwarz

but have not been tested through a life cycle of any species. Remember, only those items that have been used with some success are listed in this text, while many alternatives are possible.

Seed bugs feed with the rostrum held perpendicular to the body. The rostrum is quite thin and flexible. During feeding, the part of the labium closest to the point of attachment folds backwards during feeding so the bug can use the strength of its legs to drive its stylet into fairly solid seeds.

REPRODUCTION AND DEVELOPMENT

Seed bug mating is similar to many other true bugs. The pair mates facing opposite each other with the abdomens joined. Copulation proceeds for a few minutes to many hours. Males and females are easy to differentiate because males are visibly smaller and they tend to be differently marked on the underside of the abdomen. If there is a preponderance of males, they fight over females and fervently try to dislodge other males. However, the first male in place tends to be the winner. It is a good idea to remove large numbers of excess males or the females become worn-out from constant mating and die prematurely. In nature, they live in groups of equivalent numbers and can disperse over a wider area. Therefore, males are unlikely to have trouble locating a female and mating and fighting are much less intense.

Eggs are tiny (about 1/30", < 1 mm) and cylindrical to spherical in shape. The ova have a hard, thin shell. The shell is see-through but the egg appears to be golden yellow to brown or red and can change as the nymph inside develops. The eggs shrivel up under particularly dry conditions because the shell is permeable

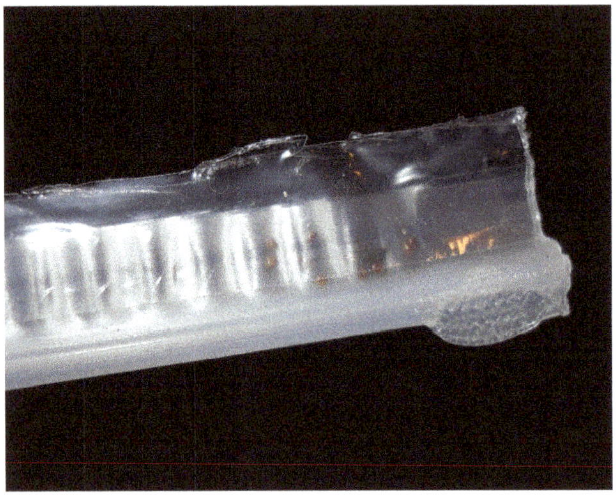

Jadera eggs are laid in crevices. This egg-laying contraption is cut from a disposable plastic zipper bag.

to water and air. Even if the eggs appear "sucked-in," they are durable and usually still hatch when moistened if the problem is noticed quickly—more than a few days and the eggs are likely dead. Drying up of the eggs severely lengthens hatching time. Eggs hatch after one to twelve weeks depending on species, temperature, and humidity.

Adult females deposit the eggs in one of two preferred sites. Any cottony substance, such as a cotton ball, folded cheese cloth, or polyester fiberfill is the preferred egg-laying site for many species (e.g. *O. fasciatus,* the small milkweed bug *Lygaeus kalmii* and the cotton stainer *Dysdercus suturellus*). Other species (e.g. boxelder bug *Leptocoris trivittatus* and *Jadera* spp.) prefer to deposit eggs under objects, such as a small piece of bark, or within cracks. Species that place eggs under objects tend to produce only one egg at a time while those seed bugs that oviposit in fibrous materials deposit ten to thirty eggs in a cluster. Although some eggs appear slightly sticky and clump together, the clump falls apart with a touch. Seed bugs generally do not glue eggs to objects, unlike leaf-footed bugs and pentatomids. It is necessary to remove eggs from the parents' cage if they cannot be adequately shielded from cannibalism. Some seed bugs, including *L. kalmii*, will suck dry every exposed egg.

Small nymphs of nearly all species are brightly colored a warning red. At each molt the nymphs grow closer to the adult coloration. Even brown-colored species customarily have bright red hatchlings. Nymphs molt once every week or two until the fifth and final molt, which requires two to four weeks. Unlike many larger insects, even the last molt is successfully achieved on the smooth sides of a plastic or glass cage (provision for molting surfaces is therefore unnecessary). Upon reaching adulthood they continue to live from six months to two years. In temperate climates, most seed bug species overwinter during the adult stage.

HOUSING

In the wild, staying together protects seed bugs from being eaten by animals as the group presents a large colorful warning of their distaste.

Some of the larger plant bugs like this *Lopidea* sp. may resemble seed bugs, but are smaller and normally have a humped appearance to the wings (when viewed from the side). They lack ocelli like pyrrhocorids, specimen in Montreal, Canada. © Martin Gamache

Small milkweed bug, Montreal. © Martin Gamache

Lopidea sp. nymph in Montreal. © Martin Gamache

round caging. The flat-sided bowl's shape directs waste droplets towards the curved, nonviewing sides. Standard screen cages can prevent the viewing problem but allow small nymphs to escape. Fabric caging is rapidly fouled and it is difficult to clean off frass. Furthermore, heavily vented enclosures lead to drying out of the animals, which tends to exceed even the adult bug's ability to obtain moisture through drinking. Hatchlings have a much greater surface area to volume ratio than adults and can dry out rapidly.

Spilostethus pandurus in Morocco. (cc) Gail Hampshire

LEAF-FOOTED BUGS (COREIDAE)

The Family Coreidae consists of medium to large bugs that somewhat resemble the seed bugs. The namesake genus *Coreus* comes from the Greek word for bug, *koris*, though the Latin word *corium* means copper. This may refer to copper highlights seen on *Coreus* spp. While a number of species have ordinary, if enlarged rear legs, members are most commonly known as leaf-footed bugs since there are various flashy species with leaf-shaped rear legs. Coreids may be closest to the seed bugs and likewise possess a four-segmented rostrum and four segmented antennae. On the other hand, leaf-footed bugs are commonly mistaken for assassin bugs by the general public. The body is an elongated oval shape and the scutellum covers only about ten percent of the abdomen. The rostrum is often shorter and thicker than the seed bug rostrum and may not extend past the thoracic segments. The rostrum can be notably truncate; and in *Acanthocephala, Thasus,* and a few other genera, it is only about thirty percent of the length of the body. The antennae are mostly large and conspicuous and can include large, flattened segments. Coreids have thick, leathery hemelytra (the family has been referred to as 'leather bugs' in some writings) and a large number of veins in the hemelytra. Additionally, the sides of the abdomen flare out and are visible on each side of the closed wings. The wings of seed bugs normally cover the sides

Leptoglossus adults mating (female is the large one), Sacramento County, California. Members of this genus are found commonly across most of the United States. They often invade houses in the fall while looking for hibernation spots. Many people mistakenly identify members of this genus as assassin bugs and a few swear they have been bitten by them, which is not possible. There are a dozen similar species of *Leptoglossus* found across North America that can be differentiated primarily by the shape of the white band across the wings. © Jordan Landerman

of the abdomen. Leaf-footed bugs are easily differentiated from most other bugs at a glance because the back legs are commonly enlarged, or flattened and widened. The rear legs of female coreids tend to be less exaggerated or of different morphology. The most widely accepted name for this family is a bit of a misnomer. The foot (tarsus) is of the usual variety; it is the lower leg (tibia) that is shaped like a small leaf.

Acanthocepha granulosa adult male.

Many of the 2,300 leaf-footed bug species (Cassis and Gross 2002) are among the most extravagant of insects (150 of these are found in the U.S.). The sides of the abdomen are often decorated with bright and dark alternating colors and sometimes further outfitted with spines or lobes. Even the antennae are often colorfully patterned in black with orange or red tips and further decorated with leaf-like expansions. Many popular entomology texts that include bugs have a disproportionately high percentage of leaf-footed bug photographs because they are so impressive in appearance. The most widely publicized are members of the genus *Anisoscelis*. They possess gigantic and colorful rear tibia flattened and widened into a leafy shape and each one of these segments is nearly as wide and long as the entire bug itself. *Anisoscelis* spp. are found from Brownsville, Texas, in the U.S., down through tropical South America. Common leaf-footed bugs in the genus *Leptoglossus* have leafy tibia, but not nearly so large. *Leptoglossus* are found most widely in Europe and the Americas, with more variety (not more quantity) found in warmer areas (nine species occur in subtropical Florida). The southwestern mesquite bug *Thasus neocalifornicus* is easily the most massive and spectacular terrestrial bug in North America, though its strange rear legs are not leaf-like. Males are armed with massive hind femora decorated in bold black and red and fitted with huge spines. *Acanthocephala femorata* males are also armed with huge, single-spined hind femora while the females' rear tibiae are leaf-shaped. The males of the majority of coreid species possess these oversized, spined femurs to fight over females, especially large animals like *Anaplocnemis* spp. from Africa and Asia. Asian *Derepteryx* spp. have leaf-like rear tibiae, but also massive,

Acanthocepha granulosa subadult female.

Adult male *Mozena* spp. from the southwestern U.S., Arizona.

leafy expansions of the pronotum resembling outfits from old science fiction flicks.

Coreid nymphs are often large and showy and can be even more spectacularly armed and colored than the adult. *Acanthocephala femorata* hatchlings begin as a see-through yellow with red markings and possess serrated abdomens; after the first molt, they change to a metallic bluish black color. Nymphs of an unidentified *Acanthocephala* species from the northeastern U.S. start out blood red and are covered liberally with gigantic black spines. Immatures can be differently colored than the adults. Nymphs of *Mozena* spp. hatch out red, turn to green after a molt, and change over to brown after reaching maturity. *Thasus neocalifornicus* nymphs are a deep red base color with multiple white markings, while the adults are mostly black with red and yellow markings.

DIETARY REQUIREMENTS

Coreidae, like other sap and seed feeders, feed with the rostrum held perpendicular to the body. The leaf-footed bug's rostrum is lightly thickened and best adapted for piercing soft plant tissues rather than hard seeds. The labium does not fold back near the point of attachment to facilitate the stylet piercing solid food.

Unfortunately, many of the more spectacular coreid bugs are not nearly as easy to feed as seed bugs. Although most of them will graze on seeds, they are particular in their needs and a majority require the sap of specific plant families. Adapting leaf-footed bug colonies to easily available foods can be incredibly difficult and often leads to extremely high or total die offs. Even though giant mesquite bugs can be fed fresh yucca pods or other soft seeds, they rarely survive a few days if not also offered live mesquite trees from which to drink sap. Other members of the tribe Nematopini also require live mesquite or related plants. Even when offered mesquite sap in captivity from a potted plant, adults seldom survive longer than a few months. *Acanthocephala* species are more catholic in their tastes, but can still be difficult to keep alive for more than three or four months. The eastern leaf-footed bugs (*Leptoglossus phyllopus*) are easy to keep alive for long periods at cool temperatures but the nymphs require a constant supply of fresh food or they fall over dead.

Acanthocephala and *Leptoglossus* can survive on a diet of grape halves (flesh should be open), along with fresh green bean pods (*Phaseolus vulgaris*) and corn. Other fruits including papaya, mango, strawberry, tomatoes, and ripe banana peels have been used. The bugs attempt to feed on shelled sunflower seeds, but are unlikely to survive long when not given softer foods like those mentioned above. Nematopini including *Thasus* can be reared on

Western leaf-footed bug nymphs, Elk Grove, California. © Jordan Landerman

Leaf-footed bug, Tucson, Arizona. © Martin Gamache

Adult *Thasus neocalifornicus*, male (left), female (right).

Leptoglossus zonatus eggs hatching. © Jordan Landerman

small live mesquite plants plus fresh green beans and grape halves. The common and more easily kept squash bug, *Anasa tristes*, is rather plain and small for a coreid and feeds on a variety of squashes. Like most land dwelling bugs, leaf-footed bugs should be provided with a source of drinking water.

While adults may survive weeks or longer with little or no food, the growing stages are not so hardy. Nearly all the leaf-footed bugs require daily maintenance of food to survive (with exception of first instar nymphs and hibernating adults that require little to no food). Food items that appear even slightly withered are ignored or fed on sparsely. They may seem to accept lapses in care or survive without food for a time, but they will not recover when fed and will not make it to the next molt. The seriously close attention required to keep these bugs fed and alive is the reason they are rarely maintained in the laboratory and why they are never seen on display at insect zoos. The soft foods required are perishable and quickly become useless as they either dry out or mold rapidly. If live food plants are offered, care can be easier, but enclosures must be larger and include lighting to keep the plants alive.

Leptoglossus zonatus eggs (approx. 1 mm each). © Jordan Landerman

REPRODUCTION AND DEVELOPMENT

The gender of coreid adults is easy to identify; even for the species whose males do not possess exaggerated back legs. The male's abdomen is concave and the last section of the lower abdomen appears notably different. He has a hemispherical genital operculum that looks like a jockey cup. The female has a rounded abdomen ending in a wide flat plate that looks the same as nearby segments.

Although leaf-footed bug mating itself is unremarkable, battles over females can be impressive. The battles between heavily equipped males can last for hours as they grab and fight each other with their gigantic, armored hind legs. The ferocious looking spines, despite their appearance, seldom inflict damage (too many

Subadult *Mozena* spp. nymphs.

holes, though, and a male can run out of blood). Usually the male with the biggest legs wins. The male simply grapples the female, then during mating, he and the female face outward and are joined by the end of the abdomen. The pair usually spends less than an hour in conjunction and rarely moves.

The eggs are thick and durable, more so than most other bug eggs. Although the ova of different species can be variable in design (usually the shape of a helmet or half a barrel) and coloration (brown, gold, or silver), they have one thing in common: a flat side. The flattened side of the ova is securely glued to the surfaces of leaves, stems, and bark. Females deposit the eggs in short rows and columns, usually from five to twenty in a bunch. In captivity, females of most species readily glue eggs to the sides or lid of the cage. If females refuse to deposit ova on the cage itself, they will oviposit on a horizontal piece of metal window screen or thin branches of the food plant. *Acanthocephala* ova hatch in about fifteen days while *Leptoglossus* can take as little as five. *Thasus neocalifornicus* ova require three to four months. Eggs requiring a long hatching time more easily dry out in the arid atmosphere of a home or mold if too much moisture is provided. They can be placed in a vented enclosure over a screen-covered container of water.

The young grow more slowly than the preceding groups of true bugs. They require two to three weeks between each of the last four molts. The first molt requires little to no food and only takes about a week. Hatchlings are different from other true bugs because they usually begin life with the required nutrients to complete the first molt. This may be confusing to the new keeper because underfed or starved 1st instar nymphs will seem to be perfectly healthy and growing. Yet, if the second instar nymphs are not fed well they die after a few days. Usually about 10% of the young of most species will die between 2nd instar and adulthood even when perfectly cared for. Losses can be attributed to several different reasons (mismolt, lack of feeding desire, and imperfect food). Species in the tribe Nematopini, on the other hand, have a high die-off rate of over 50% even when live mesquite is included in the diet (while without live mesquite, die-off is 100%).

Captive-reared subadult *T. neocalifornicus*.

Adults of temperate species usually live a year because they hibernate. In temperate climates the adult stage overwinters. *Leptoglossus* species in the northern U.S. often come indoors in the fall looking for a place to hibernate. These common leaf-footed bugs are often considered a nuisance pest (member of this genus are very often confused for assassin bugs). They survive the entire time without feeding. *Coreus marginatus* is a European counterpart with flared abdomen and standard legs that also overwinters. In arid to tropical climates the egg is usually the resting stage and the adult stage may not live longer than four months independent of food availability.

Nematopus indus in French Guiana. © Martin Gamache

HOUSING

As mentioned previously, the hatchlings of most species are large enough to be easily contained by standard (18 x 16 mesh) window screen. Any container with screen ventilation makes an adequate cage for a colony. Only the lid or one side needs to be screened unless ambient humidity is above 60%. For best success, the cage should be well lit and large enough to accommodate a small potted food plant of the appropriate species/family of plant. Cultures kept in cages less than two gallons (4 L) and containing ten to fifteen individuals are more successful than those in large cages or cages with numerous animals. Most species are not gregarious and it is rare to find multiple nymphs within antennae's length in nature. A few branches or a piece of metal window screen should be placed at an

Leaf-footed bug in French Guiana. © Martin Gamache

angle across the cage to offer surfaces for larger nymphs to successfully molt, as well as to provide egg laying sites. If possible, the cage should have easy access for a food dish since food items need to be replaced once a day. A water and wick container can be placed in the cage to provide moisture.

Old food should be removed every few days. It is helpful to clean cages once a month, but not during the first two to three instars. Handling small nymphs can result in their demise.

Cactus bug, *Chelinidea vittiger*, Pima County, Arizona. (cc) Katja Schulz

Giant mesquite bug nymph, *Thasus neocalifornicus*, in Arizona. (cc) Geoff Gallice

Diactor bilineatus, Ecuador. (cc) Geoff Gallice

Common gumtree shield bug (*Poecilometis* sp.) nymph shedding skin. Queensland, Australia. (cc) James Niland

Green shield bug (*Palomena prasina*) nymph, Oxford, UK. (cc) Danny Chapman

SHIELD BUGS, STINK BUGS, AND FALSE ASSASSINS (PENTATOMOIDEA)

Pentatomids are easy to recognize due to the pentagonal body shape. They have a broad body and a huge scutellum, covering from a quarter to nearly the entire abdomen. The antennae are usually five-segmented. The rostrum is long and rests back against the body. The legs are short and moderately uniform. Nevertheless, some predacious species are equipped with slightly enlarged, flattened front legs and males of some species have enlarged rear legs. Pentatomids are commonly called stink bugs because they are able to emit a strong chemical defensive odor from glands in the thorax. Worldwide there are more than 7,200 described species with only about 350 found in the continental U.S. Many of the former subfamilies (Scutelleridae, Acanthosomatidae, and Podopidae) are now raised to families. Around fifteen families are grouped together under the Superfamily Pentatomoidea. The Family Tessaratomidae, giant shield bugs, is one of the smaller families in terms of members, but they can be nearly two inches long (4.7 cm). The negro bugs (Family Thyreocoridae) and burrower bugs (Family Cydnidae) are differentiated primarily by the presence of spines on the front legs. These two families combined only contain a few hundred species and they are all tiny in stature (just a few millimeters in length). The various Pentatomoidea have an enlarged scutellum, a five-segmented rostrum, the same defensive secretion, and a broad body shape.

Shield bug and false assassin bug are just two popular names for members of the Superfamily Pentatomoidea. This group includes a great variety of giant, colorful, and well-armed bugs. A considerable percentage of species possess large spines jutting out from each side of the pronotum while some have spines on the abdomen or scutellum. Some tropical species have strangely elongate abdomens. Not only

Scutelleridae are among the most colorful and largest shield bugs. Nymphs of this group do not have a large scutellum until the final molt.

Green subadult *Chinavia hilaris* nymph.

This is a relative of the man-faced bug, a *Catacanthus* sp.,
Mindanao Island, Phillipines. © Christian Schwarz

Adult *Halyomorpha halys*, female found in Berea, Ohio.

Adult *Halyomorpha halys*, male found in Berea, Ohio.

Halyomorpha halys nymphs are often solitary or found in groups of only two or three by 3rd instar, Brunswick, Ohio.

Euthyrhyncus floridanus adult male, Alabama stock. © Henry Kohler

Sexual dimorphism, pentatomid females are usually easy to tell from males at a glance when both are present.

Euthyrhynchus face. © Henry Kohler

are many shield bugs beautifully pigmented with bright or metallic colors, a large number of species also have two or more color phases. Offspring are commonly different colors at different instars. In some species different offspring from the same parents may grow to become black and red, blue and orange, green and blue, etc. European green stink bugs, *Nezara viridula*, display two color phases in the same adult animal. They molt out green following the ultimate molt and change to brown in the fall to overwinter. They change back to green in spring.

False assassins are not equipped with the demobilizing venom, nor the raptorial, hairy, or sticky front legs of the true assassin bugs (Family Reduviidae). Due to some similarities in body design and diet, a few popular insect books contain photographs of predatory pentatomids inadvertently labeled as true assassin bugs feeding on other insects. Although these photographs do not always reveal the greatly enlarged scutellum, they do depict feeding and the forward projecting, straight rostrum.

DIETARY REQUIREMENTS

Some shield bugs feed only on plant juices, and some are compulsory carnivores, but the majority falls somewhere in between. Typical omnivorous shield bugs include the green stink bug (*Chanavia hilaris*) and the rough stink bug (*Brochymena quadripustulata*). The gigantic (2" or 5 cm), metallic Hawaiian koa bug (*Coleotichus blackburniae*) feeds on the sap of koa and 'a'ali'i plants (Howarth and Mull 1992), while *Trichopelpa semivittata* reportedly feeds only on wild carrot and related plants (Arnett 1993). The man-faced bug (*Catacanthus incarnatus*) feeds on the flame tree (*Delonix regia*) but also on various fruits and cashews. Pentatomids that are obligatory predators of

Predatory pentatomids simply tap into the side of a large prey. Nearby nymphs are alerted and congregate on prey (a large, teneral *Nauphoeta cinerea*).

After a few hours the abdomens of the nymphs have swolen and the prey has been emptied out.

Predation, *Euthyrhyncus floridanus* feeding on a 2nd instar hissing cockroach nymph.

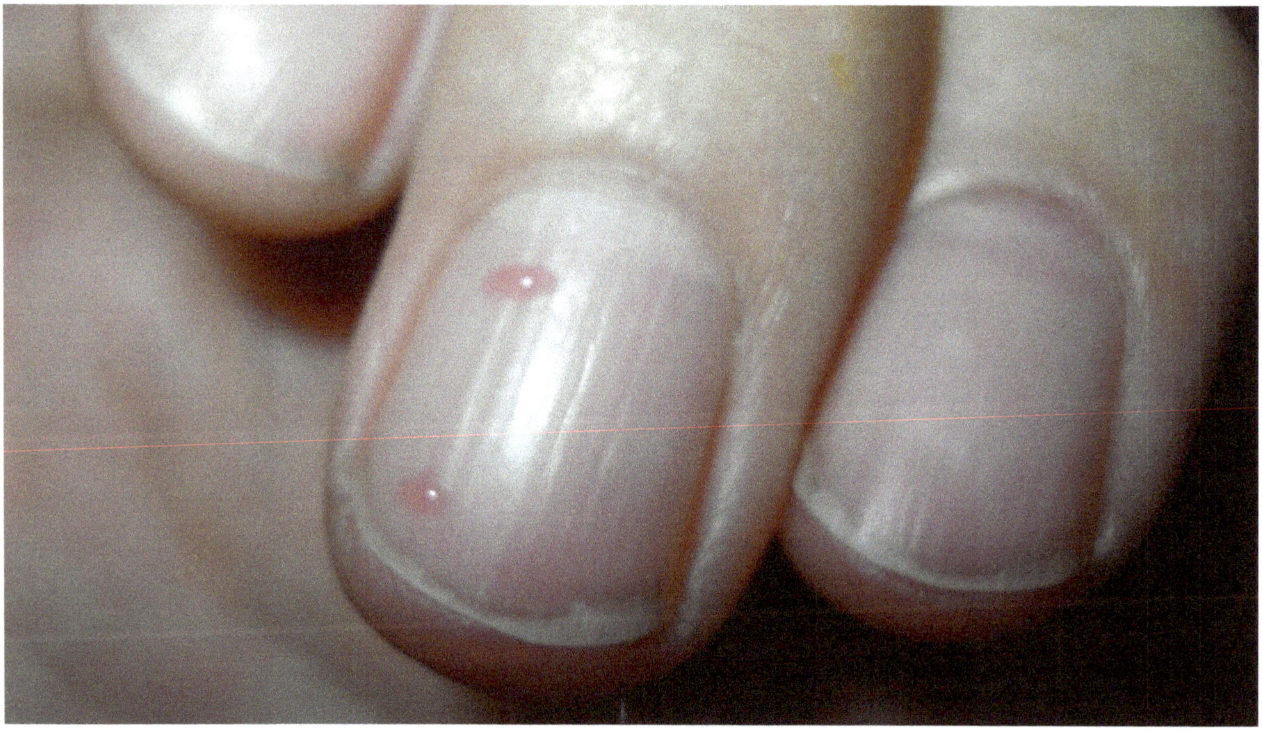

Defensive fluid droplets.

Feeding on a large scarab larvae.

other living arthropods may be called false assassin bugs or soldier bugs. Some outstanding North American examples include colorful members of the genera *Euthyrhyncus* and *Stiretrus*, as well as the rather drab *Podisus*. Although an extremely rare occurrence, herbivorous, omnivorous, and predacious pentatomids can bite people. There is no poison involved and the discomfort only lasts a few seconds.

Only a small percentage of the Pentatomidae have a negative effect on man—others are beneficial. Some herbivorous species can damage or cause distaste to certain crops, while predatory species are beneficial when they feed heavily on a variety of crop-damaging insects. The marmorated stink bug (*Halyomorpha halys*) from China has become established across portions of Europe and the eastern U.S. in the last decade with much negative fanfare. It feeds on a wide variety of fruits and vegetables, but is mostly familiar because it overwinters in buildings. Adult *H. halys* are commonly encountered walking around homes in the dead of winter. The colorful harlequin bug, *Murgantia histrionica*, is an infamous North American pest of cabbage. The two-spotted stink bug (*Perrilus bioculatus*) is a ravenous predator on the eggs, larvae, and adults of the pestiferous Colorado potato beetle (*Leptinotarsa decemlineata*), as well as various other crop-damaging beetles, Lepidoptera, and other Hemiptera.

The Pentatomid bug rostrum is adaptable to different kinds of foods. When hardened foods like dry seeds are being consumed, the labium is folded back near the base, held perpendicular to the body, and the stylet is pushed into the food. For softer foods, including fresh green bean pods and other insects, the entire rostrum can be positioned straight forward, upward, or straight down. When a bug is sucking the life out of another arthropod, the rostrum is held forward from the head, or used to hold the entire prey item in the air.

Omnivorous and predatory shield bugs are easy to feed in captivity. As mentioned above, most of the Pentatomidae are omnivores, therefore, they accept a variety of foods in captivity. Adapting a culture to easily available foods seldom causes noticeable die off. Omnivores such as green stink bugs (*Chinavia hilaris*) feed on seeds (cracked-corn, shelled-sunflower, soy beans, snap peas, peanuts, etc.) and fruits (raspberry, blackberry, strawberry, etc.) and like the occasional fresh dead cricket. Insect eggs, including moth and *Acanthocephala* eggs, are a favorite food of green stink bug nymphs. False assassins such as *Perilus biocutalus* do well when fed mealworms, cockroaches, crickets, or silkworms, but will also suck moisture from a grape. However, predatory species require prey and do not survive for long if fed only grapes. Extra care must be utilized if live crickets are being used as food because they will devour the false assassin's eggs and molting nymphs. Cockroaches are the most

Drinking from a fruit jelly, *Euthyrhyncus floridanus*.

Feeding on a roasted soybean, *Chinavia hilaris*.

Notice the rostrum stylet separation when feeding on a peanut, *Chinavia hilaris*.

Clustering behavior, 2nd instar *Euthyrhynchus* nymphs.

Molting to 3rd instar.

Euchistus tristigmus mating pair, Brunswick, Ohio.

2nd instar nymphs are raiding eggs of newly hatched nymphs, *Euchistus tristigmus*. Only seven of fourteen made it out alive. So-called plant eating species are more likely to cannibalize other broods.

Euchistus empty eggshells.

commonly employed prey in culture because they can be chosen for size, are easily cultured, and are unlikely to turn the tables on their predators (whereas crickets eat molting bugs).

False assassin bug hunting strategy is very different from that of true assassin bugs. They do not overpower or hold onto prey with the front legs, nor immobilize it with venom. They simply tap into the side of their quarry and either lift it into the air or follow it around until it dies. Also, unlike the true assassin bugs, false assassin nymphs customarily work together to bring down large prey and feed together cooperatively. This gregarious behavior can also serve to protect them from being eaten by their predators. A large colorful group advertises distaste.

Eggs from *Chinavia hilaris* on a staghorn sumac leaf (laboratory oviposition).

REPRODUCTION AND DEVELOPMENT

Males are attracted to pheromones the females release. In captivity, mating and courtship behaviors are similar to that of seed bugs, except that the males never seem to fight over females. They may flap their wings to obtain the female's attention once located on her back. Males are usually ten to thirty percent smaller than females and often have a different color pattern

Mating *Chinavria hilaris.*

on the underside of the abdomen. Even captive-reared adults display distinct sexual dimorphism in size. The last segment of the adult male's abdomen is not visible in certain families/subfamilies. As a further distinction, the male's wings cover the entire abdomen while the female's abdomen often sticks out narrowly on either side of the folded hemelytra.

Pentatomid females glue their eggs to the leaves of living plants and bark in loosely arranged clusters. Each grouping of eggs contains a dozen to more than fifty ova. Species that naturally deposit eggs on living leaves are willing to deposit eggs on pieces of metal window screen in captivity and may lay eggs on screen or metal surfaces of the enclosure. Eggs laid on

Euchistus tristigmus nymphs molting to 2nd instar. 1st instar shield bugs often do not feed.

Captive-reared *Chinavia hilaris* with her eggs.

living leaves will be less susceptible to desiccation. Eggs are prone to drying out and should be misted daily or kept over a screened water dish. Any water sprayed on the eggs should dry up in an hour or they should not be spritzed. Misting may need to occur more than once a day. Inadequate moisture is indicated by indents in the egg walls, uneven development (only a few eggs change color), and eggs that fail to hatch.

Molt from the colorful nymph to a green adult.

The ova are usually barrel-shaped, often with a row of small spines around the operculum. Eggs come in a variety of sizes (usually less than 1 mm across) and colors according to the species or developmental stage. The eggshell of many pentatomid eggs (like members of other bug families) is clear, so coloration is the result of the yolk pigmentation. The development of the nymphs inside can be observed. Transparent eggs start out light yellow in color and slowly transform to an orange or reddish-brown color before hatching. Upon close examination, the eyes and antenna can be seen as they develop and are first to darken in color. Hatching occurs after ten to twenty eight days, but may take much longer if the eggs had been

Euthyrhyncus floridanus egg clusters are usually large in number but the eggs are small for the size of the adult. The substrate is egg carton.

allowed to dry out for more than a few days. Eggs in a single cluster hatch within a few minutes to a few hours of each other.

For most species, adult specimens do not eat eggs of their own species. However, nymphs cannibalize eggs or, if the shells are too thick, wait around to cannibalize hatchlings. Therefore, egg clusters formed at different times should be moved to individual enclosures.

Shield bug aggregations tend to be much more specific and organized than those of seed bugs. After hatching, nymphs may stand around the empty eggshells in a single or double circle. All the hatchlings face either inwards or outwards from the circle. Natural groupings consists of similar aged individuals, nearly always from the same egg cluster, and does not include adults. As the immature bugs grow older, groups become less organized, but are still cohesive unless disturbed. Once the disturbance has passed they come back together. Herbivores usually cluster in large groups during the 1st and 2nd instars, while 3rd and 4th instar nymphs are found in groups of two to three animals. Predacious nymphs

The colorful *Chinavia hilaris* nymph is far more conspicuous than the bland, leaf-green parents.

A handsome *Chinavia hilaris* nymph.

This metallic beauty appears to be a Scutelleridae but the antennae are four-segmented. Panay Island, Philippines. © Christian Schwarz.

cooperate to bring down large prey and can remain in tight-knit groups through 4th instar. By 5th instar, nymphs of most species no longer associate closely. Adults disperse in nature and are infrequently found in couples (male and female, naturally). In artificial enclosures the predacious species maintain clustering through

Adults false assassins feeding together on a large cockroach. Captive-bred *Euthyrhynchus floridanus*, central Georgia stock, Jan. 2000.

5th instar and even somewhat after reaching adulthood. However, adults of carnivorous species tend to eat each other if kept together in close quarters, particularly if they are not provided with adequate moisture and food.

The nymphs molt five times, about once every two weeks. The final molt requires about four weeks. Total time from 1st instar to adult is about 90 days. As with many hemimetabolous insects, the transformation to adulthood includes the formation of wings, but also includes changes in color and shape. Nymphs are usually round or tear-shaped. The transformation to adulthood includes attainment of the classic pentagonal pentatomid shape.

Clustering behavior, 2nd instar *Euthyrhynchus* nymphs.

HOUSING

Omnivores and herbivores are normally kept together and do well even if somewhat crowded. For these shield bug species it is usually most feasible to keep adults and young in the same enclosure. Nymphs of the predacious false assassins are kept together and seem more communal. They band together even if a few molts apart or from different parents. Crowding is not a problem. However, predacious nymphs should not be kept with the adults or the young may be eaten. It is important to keep the density of adults low and food readily available. A quart (liter) of space for every three or four adult false assassins is necessary or they may kill each other.

Pentatomids, as with nearly all terrestrial Hemiptera, require a minimum amount of ventilation. The enclosure should have at least one 1" x 1" (2.5 cm x 2.5 cm) hole covered with fine screen or cloth. Bugs cannot chew through any material, though prey such as crickets can. There is no maximum amount of ventilation

Shield bug, Montreal. © Martin Gamache

Shield bug in Argentina. © Martin Gamache

Brochymena affinis, Montreal.
© Martin Gamache

A predatory nymph, Subfamily Asopinae,
Montreal. © Martin Gamache

Edessa rufomarginata, a widespread
plant-feeding species in South America.
Adult in French Guiana. © Martin Gamache

Banasa dimidiata, Montreal.
© Martin Gamache

but full-screen enclosures are generally too large or have cracks immature bugs can escape through. Also, eggs and food can dry excessively calling for constant misting and replacement. Hatchlings of even the largest North American species are just small enough to fit through ordinary window screen; therefore, microscreen or cloth ventilation is necessary for containment. Hatchlings of monstrous tropical shield bugs like the man-faced bug or koa bug start out large, so ordinary (1 mm x 1 mm) window screen would be sufficient.

Hatching *Euthyrhynchus floridanus*, 27 days after oviposition.

Shield bugs can be kept in the same type of containers as seed bugs or leaf-footed bugs. Habitats should not exceed 32 oz (950 ml) or small nymphs will be difficult to feed and locate. Inside the cage, a piece of metal window screen should be provided for molting and ovipositing. The screen should be placed vertically or at an angle. A piece of egg carton can be placed in the cage to help prevent young from hanging out in the corners of the lid where they are easily crushed during feeding or cleaning activities. Even the most gregarious species tend to run rapidly in all directions when disturbed.

Foods, especially those provided to omnivores, rarely contain adequate moisture, so it is important to place a wick and water container in the cage. Humidity will reduce the amount of fluid required but bugs cannot drink humidity. Cages need to be cleaned only once every month or two. If old food builds up, and the decomposing pile appears wet, it should be removed from the cage immediately.

Metallic shield bug, *Scutiphora pedicellata*, Australia. (cc) John Tann

ASSASSIN BUGS (REDUVIIDAE)

ASSASSIN BUGS are all members of the family Reduviidae, though not all reduviids are assassin bugs. Conenoses that feed on vertebrate blood are not assassins, as the term is reserved for the predatory groups. Predatory reduviids are easy to identify by their unique, curved, and short three-segmented rostrum. No other terrestrial bugs possess a similar appearing mouth. The rostrum curves backwards below the head and tends to remain pointed backwards, even during feeding. Most of the seven thousand Reduviidae species are found in tropical areas. The U.S. boasts 184 species, primarily found in the South. In most habitats the largest, most impressive, and most common species are from the subfamily with the most species, the Harpactorinae. The large, common American *Arilus* and the Australasian *Pristhesancus* are prime examples of this group. Most texts place the diurnal ambush bugs (subfamily Phymatinae) within the Reduviidae. These are small (about 1/3-inch or 0.8 cm) colorful species that await prey in flowers like the generally larger *Apiomeris* spp. The longest U.S. species are the thread-legged bugs (subfamily Emesinae). *Emesaya brevipennis* can exceed 1.5″ (nearly 4 cm) and is found across North America.

Group of *Platymeris* sp. "Mombo" 2002.

The reduviids possess the most variable design and body shape of the true bug families. The front legs of assassin bugs can be thickened and raptorial, elongated and baton shaped, covered with spines, or otherwise modified for use in catching prey. However, the front legs of obligatory blood feeders, as well as some insectivores, are ordinary walking legs. Assassins use conspicuous and uniquely adapted antennae to touch and poke objects to decide if such objects are worth tackling with the raptorial legs. The abdomens of many reduviids stick out past the sides of the wings and may be brightly banded in color, similar to the Coreidae. The scutellum may be unadorned, or decked out with large spines like *Psytalla*

Notice the exaggerated thoracic crest of the wheel bug (*Arilus cristatus*).

Repipta taurus captive-reared male, Homestead, Florida. © Alan Jeon

Pseudozelurus arizonicus male, a ~2 cm species endemic to Arizona, USA. © Alan Jeon

Pnirontis languida collected under lights at a tennis court in Montgomery, Alabama. © Alan Jeon

Psytalla nymph eating an adult *S. lampyridiformis*.

2nd instar *P. rhadamanthus* feeding on an adult *S. lampyridiformis*.

Arilus cristatus shortly after the molt has numerous, small, long setae but lacks the tiny, gray structures. © Henry Kohler

Arilus cristatus nymph eating a *Drosophila melanogaster*. © Henry Kohler

Rostrum strength hisser and Mombo. *Platymeris* sp. "Mombo" feeding on an adult male hissing cockroach. The female assassin is holding (and touching) the roach with her rostrum only.

assassin bug in the world in terms of size, color, and armature. I first saw photos of specimens in the mid 1990s (though at least one photo was a brightly colored, mislabeled Harpactorinae). Specimens began to show up as laboratory-cultured a few years ago. Hopefully in the not too distant future there will be more research and interest in these spectacular bugs. More than a couple of species should be readily available to those interested in studying these bugs in captivity.

DIETARY REQUIREMENTS

Assassin bugs are so named for their excellent killing abilities. Insectivorous reduviids kill and consume a wide variety of creatures including ants, spiders, grasshoppers, snails, millipedes, centipedes and just about every terrestrial invertebrate that exists in nature. Many species have been investigated for their use in biological pest control. Nocturnal assassins like *Platymeris* and *Microtomus* feed on ground-dwelling cockroaches, crickets, and other nocturnal crawling insects. Diurnal Harpactorinae tend to feed on day-active insects. The large Australian Harpactorine *Pristhesancus plagipennis* sits on flowering plants where it feeds on hymenoptera, butterflies, buprestid beetles, leafhoppers, grasshoppers, and flies (Hawkeswood 1992).

Crickets, mealworms, fruit flies, sowbugs and pillbugs (terrestrial isopods are crustaceans, not insects, let alone true bugs), seed bugs, waxworms, and cockroaches (especially climbing species like *Schultesia lampyridiformis* and *Blatta lateralis*) are commonly

Gardena sp. (Emesinae) male hunting on a web. "This species is almost exclusively associated with spider webs. It feeds on the prey caught in the webs, and the spiders themselves. The male has slightly longer wings than the female." Pers. comm. and © Christian Schwarz

Zelus luridus captive-reared adult, Brunswick, Ohio.

Zelus luridus nymph. The sticky baton-shaped front legs nymph can be held back, ready to strike.

used as prey for cultured assassin bugs. Terrestrial isopods are the best food for many voracious species because they are readily eaten and they do not attack molting assassin bug nymphs. Mealworms are readily accepted by the more timid species and, unlike nearly all other large prey items, they do not eat egg clusters. Although other predatory hemipterans (especially the aquatics) are adamant scavengers, assassin bugs cannot be reared on frozen or dead insects. However, extremely hungry reduviids may feed on a dead insect if it is moved around to appear alive or if another assassin bug is already feeding on the dead prey. Unlike false assassin bugs, assassins are not likely to pierce grapes or other juicy fruit to acquire moisture even if they are very thirsty. Moisture is best provided in a shallow, sponge-filled dish.

Fruit flies (*Drosophila melanogaster*) are useful prey for small Reduviidae. Cultures are easy to set up but exhaust after month or so. A 1/4 cup water, 1/4 cup instant potato flakes and a pinch of yeast is a useful media for rearing fruit fly larvae.

Although all assassin bugs are aggressive predators, some are more aggressive than others. Hungry *Platymeris* adults and large nymphs will jump on and attack a massive, heavily-armored, adult Madagascar hissing cockroach (*Gromphadorhina portentosa*) and

Playtmeris rhadamanthus cannibalism is not unusual, but does not prevent their communal rearing.

ride the violently thrashing giant roach around until it is successfully subdued. Species of *Arilus* cautiously attack and eat prey as large as themselves, but not much larger unless it is a moth. Predation varies by the type of prey (i.e. larger moths will be taken faster than crickets) and the prey's location in the cage. The delicate green assassin bugs (*Zelus luridus*) restrict their feeding primarily to small moths and flies. Although this species takes out massive numbers of small flying insects, it seldom eats active prey more than a quarter its size. However, specimens can feed on a relatively massive caterpillar. Depending on environmental conditions, prey type and frequency, specimens may or may not require extra moisture to drink.

Reduviidae are usually solitary in nature if not in habit. None of the species kept so far have been shown to cooperate in taking down large prey items. Nevertheless, a study has shown the millipede assassin bugs, subfamily Ectrichodiinae, feed communally at least part of the time and may even hunt in groups (Forthman and Weirauch 2012). Many species cannot be kept together as nymphs, while even

House crickets, *Acheta domesticus*, are readily available prey items. © Henry Kohler

Blaberus craniifer nymphs are easily cultured prey used to feed large assassins like *Platymeris rhadamanthus*. Any of the feeder cockroaches can be reared on dry dog food and fruit.

Platymeris rhadamanthus eating a subadult lobster roach, *Nauphoeta cinerea*. This species often escapes predation and can offer some value in keeping the cage clean.

Rostrum strength as seen in this firefly mimic cockroach (*Schultesia lampyridiformis*) held aloft by a small *P. rhadamanthus*.

These Cameroon millipede assassins refuse to feed on mealworm larvae despite the prey`s similarity to millipeds in body shape. © Peter Grabowitz

Freshly molted female *M. purcis* with its prey, a *Parcoblatta lata* nymph. © Alan Jeon

communally kept species tend to cannibalize. In captivity, nymphs and adults of communally maintained species may feed together on large prey items, but they fight, sometimes viciously, over small prey. If prey is small enough it is usually carried away or held aloft so the assassin can feed in peace.

Cannibalism is not a problem among mature adults of the same species. In addition, nymphs of the same size may not kill each other when fed regularly. If nymphs of different sizes and adults are all kept together, deaths due to cannibalism can be very high if the colony is underfed. Some species are more likely to kill each other. Wheel bugs (*Arilus cristatus*) seldom cannibalize. *Platymeris rhadamanthus* resort to cannibalism generally just when really hungry. *Microtomus purcis, Zelus luridus*, *Melanolestes picipes*, and *Platymeris biguttata* nymphs can kill each other no matter how well fed they are. Feeding frequency, species, and type of prey must be considered if kept communally.

These Mombo are each feeding on their own house cricket (*Acheta domesticus*). Crickets can be reared on chicken feed or cricket chow, but are usually purchased as feeders from pet stores or cricket farms.

Though members of the other subfamilies are exclusively predators, members of the Subfamily Triatominae, known variously as kissing bugs or conenoses feed on vertebrate blood. They are considered parasites because they do not kill their prey. The masked hunter (*Reduvius personatus*, from the insectivorous subfamily Reduviinae) is sometimes confused with kissing bugs since it is found in homes. However, the masked hunter feeds on small insects, especially fleas, museum beetles, bed bugs, and conenoses that are associated with human dwellings. The frighteningly large conenoses feed exclusively on vertebrate blood. The food source is often rodents and reptiles, but large mammals are not exempt. Specimens kept in captivity refuse to feed on invertebrates of any

Ectrichodia group of adults from Cameroon. © Peter Grabowitz

Ectrichodia captive hatched immatures, the eggs were not seen but probably dropped in the substrate. © Peter Grabowitz

Freshly molted 4th instar *Apiomerus crassipes*. © Alan Jeon

Female *Apiomerus crassipes* collected from a flower (*Daucus carota*, Queen Anne`s lace) in Montgomery, Alabama. © Alan Jeon

kind. While predatory assassins seldom recover after a few weeks without food, a large *Triatoma* nymph can remain healthy for six months in a container without food or water (pers. obs.). Laboratory rearing involves setting up an artificial feeder. Reisenman et al. 2011 employed ". . . bovine blood, containing 2% sodium citrate as an anticoagulant and supplemented with 100 mm adenosine 5′-tryphosphate disodium salt (Calbiochem, La Jolla, CA) . . ."

Even the predatory assassins can be restricted to specific types of invertebrate prey. There is one rather large subfamily, the Ectrichodiinae (with 660 plus species (Chlond 2010)) that specializes in eating diplopods. These reduviids are commonly known as millipede assassin bugs because they will only feed on millipeds (often much larger than themselves). They are capable of going many months without food, which may be the result of eating large, seasonally available prey. Attempts at feeding them mealworms (which a human would imagine looks an awful lot like a milliped) have resulted in failure. I purchased a few large (30 mm), glossy, red-and-black, adult millipede assassin bugs originating from Cameroon in the 1990s and they lived for half a year without feeding. I had no idea at the time they required large diplopods and tried all sorts of prey items without success. I might have steered clear had I known, since large millipedes can be rather expensive prey items. Still they could be useful to cull sickly, damaged, or deformed diplopods. Within this unique subfamily there are a few small *Rhiginia* species found in the U.S., while large species are mostly tropical and mostly Old World.

This *M. purcis* nymph is feeding on a cowpea weevil (*Callosobruchus maculatus*). Most assassin bugs and water bugs will accept this feeder beetle. Dry cowpeas are the standard media for rearing *C. maculatus*. No moisture and no care is required, but development time is long. Feeders can be harvested about around three months after seeded with initial adults (fruit fly cultures only need two weeks for production).

Phymatinae, the genital operculum clearly shows this animal is a male. © Henry Kohler

REPRODUCTION AND DEVELOPMENT

Adult male and female Reduviidae usually appear nearly identical from above. Rarely, males have full wings while females are flightless with small wingpads like nymphs. Females of all species tend to be a little longer and bulkier on average, but size and shape are unreliable methods of gender determination (if not altogether useless). There is one consistent and

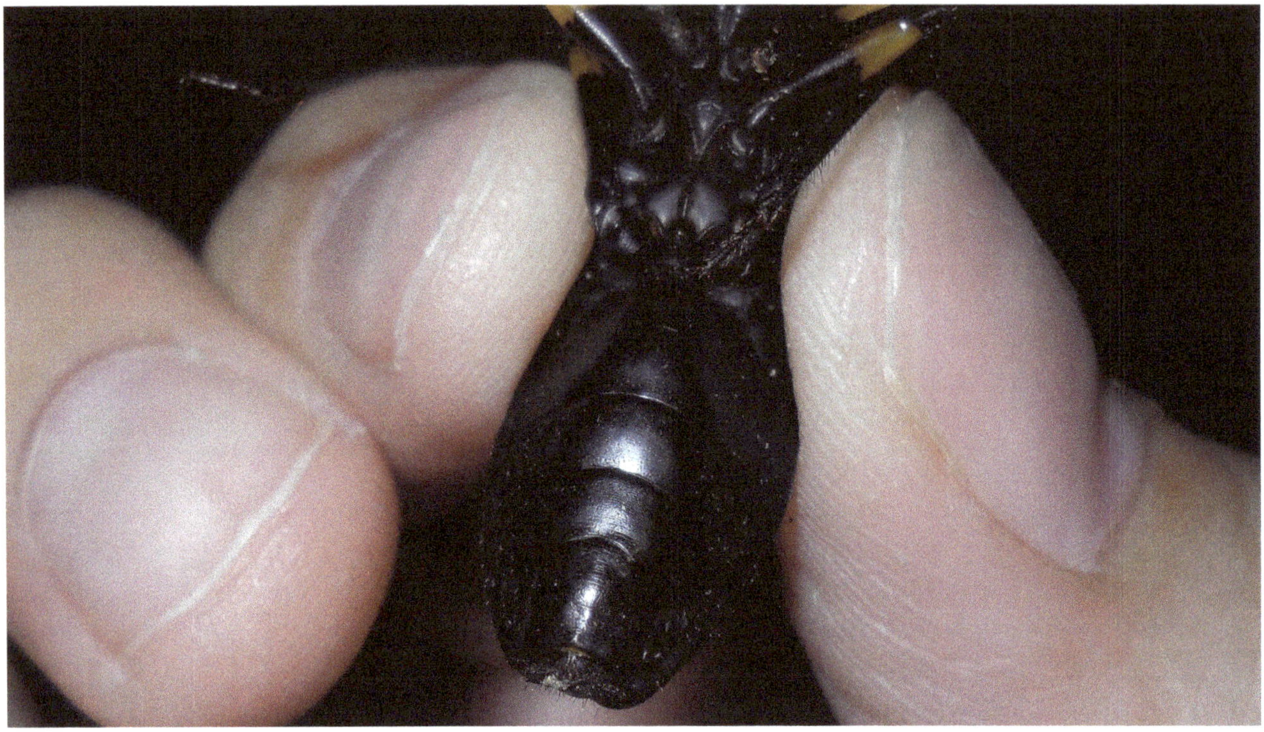

The last abdominal segment on adult females does not include a small central segment, *P. biguttatus*.

The last segment on mature males has a small central, or terminal, segment covering the male genitalia. This segment is the genital operculum, *P. rhadamanthus*.

Mating pair of *P. rhadamanthus*. Mating is not commonly noticed in Reduviidae.

Ova comparison, two small *P. rhadamanthus* and two large *P. horrida*.

easily viewed characteristic useful in identifying gender. The male's genital operculum is hemispherical and looks like a jockey cup. The female's subgenital plate is flattened and inconspicuous. Unfortunately, the difference cannot be seen on nymphs, so the gender can only be guessed at according to nuances in the size and shape of the body.

There is nothing out of the ordinary about assassin bug courtship other than it being a rare observance. Larsen (1998) states copulation only lasts for about three minutes, which is why it is rarely observed in captivity. Still, the exact time required is hard to gauge since they tend to disengage when disturbed and mating is not regularly observed. Mating is commonly side-by-side like waterscorpions or giant water bugs. Mate guarding is common behavior in the Phymatinae.

Egg laying strategies of Reduviidae fall into four primary categories. The eggs come in shapes that tend to correspond with the manner in which they are deposited.

Strategy 1: Eggs are rod-shaped and are glued to sticks or leaves in upright, compact, circular, and large (a dozen to one hundred or more ova) clusters. Adhesive material coats the sides of the eggs to make them stick together and to the site of attachment. The eggs and operculum of species that employ this strategy can appear hexagonal in cross-section, like a honeycomb, because the clusters are very compact. Clusters hatch in concert. Genera that employ this strategy include *Arilus* and *Zelus* (Subfamily Harpactorinae), and *Phymata* (Subfamily Phymatinae).

Strategy 2: Ova are egg-shaped (oval) and are simply dropped on the ground or buried shallowly in a manner similar to the approach employed by most walking sticks. Genera that employ this strategy include *Platymeris* and *Reduvius* (Subfamily Reduviinae), *Dipetalogaster* and *Triatoma* (Subfamily Triatominae), and *Melanolestes* (Subfamily Peiratinae). This and strategy 1 are the most commonly seen.

Strategy 3: Eggs are spindle-shaped and deposited singly in tight-fitting holes and crevasses (sometimes ova may to be

Ova from *P. rhadamanthus*, hatched eggs are missing their operculums.

(Top Left) Ova cluster from Phymatinae. © Henry Kohler (Bottom Left) Harpactorinae Cameroon egg cluster. Only a few from two egg clusters survived desiccation. © Peter Grabowitz. (Top Right) A hatchling and some empty shells, *Dipetalogaster maximus*. Freshly laid ova have an internal, pale pink color which darkens as the nymph develops. The shells are see-through but have a strong white cast. Triatominae egg shells are usually see-through though the operculum is still apparent.

glued lengthwise (pers. comm., Murphy)). Genera include *Microtomus* (Subfamily Microtominae).

Strategy 4: Elongate eggs are glued lengthwise, individually to the surface of bark or leaves (as in thread-legged bugs, Subfamily Emesinae).

The only somewhat uniform aspect of the eggs in various strategies is a large and conspicuous operculum. The operculum is often pigmented in white or off-white and contrasts greatly with the dark coloration of the eggs (black, gray, brown, red, etc.). Corsair eggs have a ring of outward directed filaments around the operculum, resembling eggs of the leaf-insect *Phyllium philippinicum*.

Egg clusters formed on microscreen are notoriously difficult to keep alive because the vent dries out the eggs.

It seems that eggs of species that deposit eggs singly are hardier than eggs of species that form egg clusters. *Platymeris* and *Microtomus* ova can be kept extremely dry or wet and may still hatch successfully (ova kept extremely dry do require moistening before hatching and can die or take much longer to hatch). The egg clusters of *Arilus* and *Zelus* species are decimated by dry and wet conditions. The eggs quickly mold if kept wet and desiccate if kept dry. Egg clusters should be kept in a vented container, placed over a screened water dish and misted

Small cluster of *Arilus cristatus* eggs.

daily. Take note: mistings should dry up between a half hour to one hour. If mist dries up in a few minutes or remains for hours or days, the ventilation is incorrect. Another failing of the egg cluster strategy is that in normal cages, hungry crickets, cockroaches, and even terrestrial isopods easily find and devour the entire thin-shelled egg masses. Unlike some of the herbivorous bugs, assassin bug eggs are not subject to cannibalism by adults or nymphs.

After two weeks to two months, the variably colored hatchlings push aside the operculum and emerge. Like all true bugs, five molts and five nymphal stages are required to reach adulthood. Nymphs grow slowly and only molt every few weeks. Molting is a vulnerable time when nymphs are most likely to be eaten by feeder insects and other assassin bugs. The young bugs do not eat for a few days to a week just prior to each molt. Pay attention to the nymph's appetite and do not allow live crickets to stay in the cage if they are not eaten within 24 hrs. As nymphs grow, it may be necessary to separate or thin them out. The final molt can demand up to a three-month wait. During this ultimate molt the assassin bugs usually develop flying wings and assume a different color and shape.

Although normal for aphids, parthenogenesis is rare for Heteroptera. Normally both a

Shed exoskeleton, including the linings of the trachea.

Teneral assassins are often very colorful.

male and female are a necessity for development of the eggs. Only a few Heteropteran species, including one species of water treader (Family Mesoveliidae) and one species of plant bug (Miridae) (Wheeler 2001), are known to be capable of reproducing without males. Bruins' (1999) terrarium pet guide claimed this was a means of reproduction for *Platymeris*, probably because mating is rarely observed. This has spurred rumors that parthenogenesis is a common means of reproduction for *P. biguttatus*. I have isolated females prior to maturity and later watched for their unfertilized eggs to develop with no success. At this point in time parthenogenesis has never been documented for *Platymeris*.

Notice the operculum filaments on the *Arilus cristatus* egg. © Henry Kohler

HOUSING

The artificial habitat should be well ventilated, but also must be escape-proof because assassin bugs are able to climb smooth surfaces and squeeze through small holes. Arboreal species, including bee assassins (*Apiomeris*), wheel bugs (*Arilus*), and thread-legged bugs, climb glass or plastic and prefer to perch near the upper sides or hang from the lid. Terrestrial assassins such as corsairs, *Platymeris*, and *Microtomus* are not as obvious because they usually stay on the bottom and hide among cage decorations. Terrestrial species climb glass just as well but often only climb at night. Hatchling *Platymeris* sometimes hang out in the corners of the lid and can be smashed when the lid is closed. A thin coating, two-inches (5 cm) wide, of petroleum jelly can be applied to the upper sides of the cage to curtail climbing if needed.

Entire colonies of some of the more vicious species are customarily kept together. Some nymphs may be eaten, which can help to weed out weaklings and deformed individuals. *Platymeris* and *Psytalla* nymphs hide among the decorations and only need to be reared separately if a large number of adults are desired. Die-off can be lowered if large nymphs and adults kill combative prey, while tiny nymphs feed from the huge dead carcasses. An inch or two of potting soil or wood chips is placed on the bottom of the cage and pieces of egg carton or/and wood are arranged for hiding as well as for climbing and molting. Do not forget some narrow hiding places for the young. A few small branches should also be placed in the cage for species that deposit egg clusters. A shallow, sponge-filled water dish can be placed just about anywhere in the cage. Keep the water dish full to provide drinking water for the bugs and moisture for eggs laid in nearby substrate. Egg clusters are formed on branches or the lid and should be misted every day or two. Note: extremely high nymph mortality rates are decreased by adding decorations with multiple hiding opportunities, keeping the culture well fed, and removing uneaten crickets after twenty-four hours. Crickets will attack molting nymphs and chew up eggs, except those with really hard, thick shells like *Platymeris*.

Arilus nymph molting. At this age teneral specimens are orange while teneral adults are salmon pink. © Henry Kohler

Freshly molted nymphs have short, thin abdomens that will expand many times over with feeding.

Cameroon Harpactorinae nymph. © Peter Grabowitz

Arilus, Platymeris, and *Psytalla* are commonly housed communally in ten-gallon (40 L) or larger glass terrariums with a screen lid. Prefabricated lids made with window screen work well, but reptile lids with large drilled holes or widely spaced metal mesh cannot contain nymphs. Nymphs can also escape through the large vents in pre-formed, plastic creature cages. The sponge-filled water dish should be refilled on a regular basis. Cage cleaning in the larger enclosures is not necessary on a regular basis. Large items may need to be removed to prevent phorid fly infestations.

Most other genera are housed individually in 2 oz (60 ml) vials, up to 32 oz (950 ml) clear plastic cups with microscreen vents. Terrestrial isopods and small cockroaches are a great food for rearing separated nymphs of many species. The cage should not have corners or edges that shield the food from being eaten. Arboreal species are usually provided no substrate and limited decorations. The cage is cleaned regularly to prevent carcasses from becoming the hiding place for prey. Terrestrial assassins are kept on a layer of damp substrate.

Zelus longipes is known as the milkweed assassin bug because adults resemble *Oncopeltus*. Montgomery, Alabama. © Alan Jeon

Zelus longipes nymphs do not resemble *Oncopeltus*. Lafayette, Louisiana.

Male *Zelus renardii* from Modesto, California. The prey carcass is a *Parcoblatta* sp. nymph. © Alan Jeon

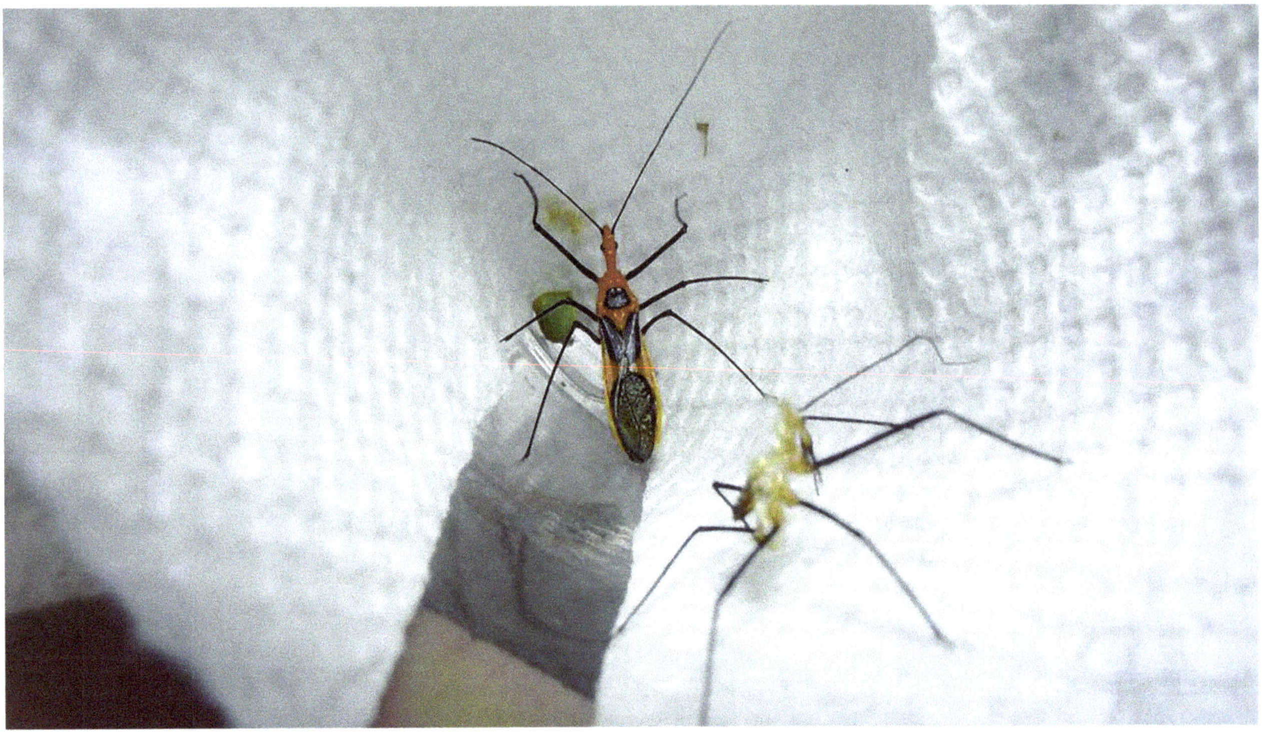

Zelus longipes freshly molted. © Alan Jeon

Check the chosen container carefully for house spiders before use. One spider can devour massive numbers of hatchling bugs before it is noticed. Assassin bugs would probably eat the spider if it did not have silk to trap them.

Temperature suggestions for rearing large tropical species, such as *Platymeris biguttatus*, are usually high during the day, 82-86° F (28-30° C) (Larsen 1998) or 79° F (26° C) (Fritzsche 2003), dropping to room temperature at night 73F (23° C). Yelich (2004) suggested 70° F with no exceedance of 90° F. Although higher temperatures do not necessarily adversely affect rearing (such as through increased desiccation), room temperature at 73° F (23° C) for the entire life cycle has proven sufficient across various genera including *Arilus*, *Emesaya*, *Microtomus*, *Platymeris*, *Psytalla*, and *Zelus*.

Harpactorinae adult from Cameroon. © Peter Grabowitz

All species are relatively easy to rear when close attention is paid to their husbandry. However, only the *Platymeris* and *Psytalla* can survive through surprisingly long bouts of neglect. This is partly due to their propensity to cannibalize under adverse conditions.

HANDLING AND VENOM

In general, the answer to whether or not the keeper or researcher should handle assassin bugs is a resounding *NO*. I have been bitten or stung by a variety of arachnids, centipedes, and insects, including a few of our native reduviids. By far, the worst pain I have experienced was from a *Platymeris biguttatus*. At the time (1998) I had been keeping this species for five years and had grown rather comfortable with handling them. When necessary for transfer, feeding, or status checks, I regularly picked up the adults by the pronotum. The width of the pronotum kept my fingers just clear of the venomous rostrum (though they very often harmlessly spat on my fingertips). I had been warned of the intense pain, but not by anyone who had actually been bitten. And yet, over time, I grew too unconcerned and one day was picking up and moving some adults by hand when I

3rd instar *Melanolestes picipes*. © Alan Jeon

Melanolestes picipes, male nymph two molts from maturity. © Alan Jeon

Black corsair (*Melanolestes picipes*) adult female, Montgomery, Alabama. © Alan Jeon

Ova operculum filaments, view from above, *Melanolestes picipes*. © Alan Jeon

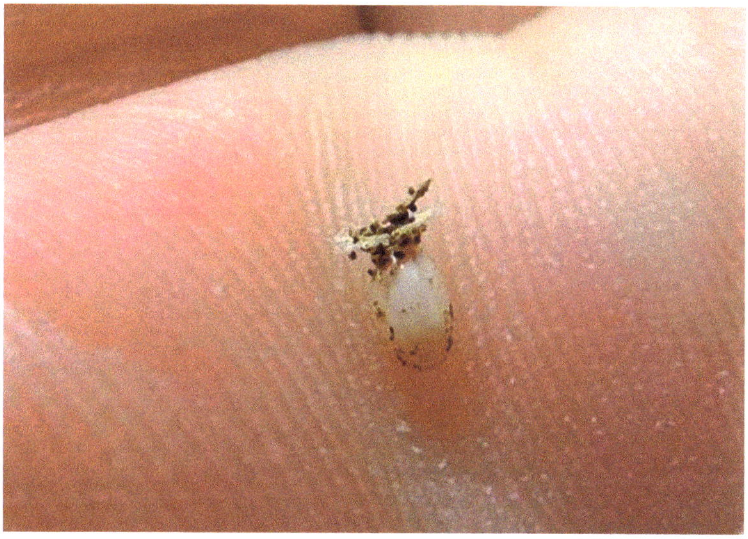

Side view of the *Melanolestes picipes* egg. © Alan Jeon

Hatchling coloration *Melanolestes picipes*. © Alan Jeon

A brightly colored assassin bug (a male) in the forest, French Guiana. © Martin Gamache

This Harpactorinae in the rainforest still hunts with amputated antennae, Suriname, South America. © Martin Gamache

grabbed a subadult by the pronotum. Of course it had no trouble injecting my thumb with a good deal of venom as the rostrum easily reached deeply into flesh. I stood in the same spot in intense pain for three hours. The bite itself, I remember, was rather painful, but not necessarily worse than certain centipede bites or bee stings. Despite faded memory, I am certain this pain was far more intense than any other envenomation. Whenever I reached into the tank (not even to pick up an adult) for nearly a year afterwards my body would shake involuntarily. No bite before or since has ever caused me such an amazing subconscious response—not one time afterwards, let alone for an entire year. Randy Morgan of Insect World Cincinnati, a keeper who had also experienced the intense pain caused by the venom of bullet ants (*Paraponera clavata*), said the *Platymeris* bite, though quite painful, was less intense than *P. clavata* (pers. comm.). The widespread masked hunter *Reduvius personatus* is said to result in "a swelling more painful than that of a bee" (Fracker 1912). "I have been warned of afrotropical reduviids encountered in the Ivory Coast, some of which are larger than *Platymeris bigutattus* (not necessarily than *Psytalla*). When I asked how intense the bite was, I was told that various men would end up crying. The only thing that seems to be worse aside from bullet ants are tarantula hawks of the genus *Pepsis*" (pers. comm., Christian Schwarz).

Unless they are pinned down, assassins do not recognize human flesh as something to bite. This makes even the most venomous seem harmless which can lead to careless handling. However, if an animal climbs onto your hand or arm the easiest response is to shake it or flick it back into the cage (not grab). As with all rules of thumb, the assassin bug propensity to never bite flesh if unprovoked is not true 100% of the time. Years ago while in my back yard I felt an intense pain in my arm. I looked down to see a half-grown *Zelus luridus* deeply ensconced in the flesh of my forearm. The pain only lasted a minute and reminded me of a water boatman bite.

Predatory assassin bug venom is designed to rapidly immobilize prey as well as begin the digestive process. The ability to quickly kill prey has been well known to naturalists for hundreds of years, such as Degeer (1773) commenting on the near instantaneous death of a fly caught by *Reduvius personatus*. Even surprisingly large prey items, like oversized tropical cockroaches and beetles, cease struggling

Mombo on hand. Although handling pets may be enjoyable, assassin bugs and some water bugs with potentially painful bites should not be held.

An adult *Platymeris* held this way may not be able to bite, but a large nymph can.

Reduviid, Vohimana Reserve, Madagascar. (cc) Frank Vassen

Assassin nymph, Texas. (cc) Josef Wells

within seconds of being bitten. This impressive chemical mechanism is the reason the front legs of assassin bugs rarely give away their predatory nature (the raptorial forelegs of Emesinae, Phymatinae, and the sticky forelegs of certain Reduviinae being among the obvious exceptions). Little attention has been paid to assassin bug venom in the literature (other than Triatominae) because it has short-lived and relatively insignificant effects on humans. Edwards (1961) performed detailed testing on the reaction of insect systems to, and the composition of, *Platymeris rhadamanthus* venom. He found it contained at least six proteins and the primary function is the breakdown of cells in external digestion rather than any specific toxicity. Furthermore, the venom proved harmless when prey insects were bathed in it or fed small amounts, so the spitting habits may be to deter vertebrate predators through a warning smell advertising the painful bite or through irritation of the mucus membranes. Edwards also found this species, at least, seemed immune to its own venom when injected into a 5th instar nymph at 0.1 ml at 1%. However the immunity could not be transferred to another invertebrate with an infusion of blood as it can be for centipede venom. This may be because *Platymeris* are not immune to their own venom at dosage rates consistent with a standard bite. Nymphs sometimes attack each other when their enclosure is disturbed. I have observed initial cannibalism on half a dozen occasions and have separated the individuals within seconds of victims being bitten. All victims have lost mobility within seconds of the bite, like prey, and none have recovered.

One very popular question scientists have been unable to answer is: "How can the assassin-bug's venom, which is very powerful and will quickly dissolve almost any insect tissue, be stored by the assassin bug for future use?" (Hutchins 1966). I have a simple answer. The venom is not that powerful and easily stored in the salivary glands. Breakdown of the prey's content is more mechanical than chemical. Like belostomatids, the reduviid stylet moves from the rostrum like a drain auger inside the body of the prey.

Baja conenose bugs, *Dipetalogaster maximus*, maintained in the laboratory. This species feeds on the blood of lizards in Baja California Sur, Mexico.

The blood-feeding Triatominae, or conenoses, have a different sort of venom that does not kill insects. Triatominae will not bite insects and their venom does not harm insects when injected artificially (Edwards 1961). This venom is not meant to be painful, nor directly harmful to vertebrate circulatory or nervous systems, because the design is to feed undetected and return for another meal. However, feeding by conenoses can offer significant after-effects and may even result in severe allergic reactions. The bite itself causes little to no pain, but after a few hours can result in painful itching and swelling. Some people develop allergies to the bites over time. *Trypanosoma cruzi* is the protozoan responsible for American trypanosomiasis or Chagas disease, a disease restricted to the tropical New World. A handful of species of *Triatoma*, *Rhodnius*, and *Panstrongylus* are the common vectors.

A wheel bug everts its brightly colored, offensive odor gland.

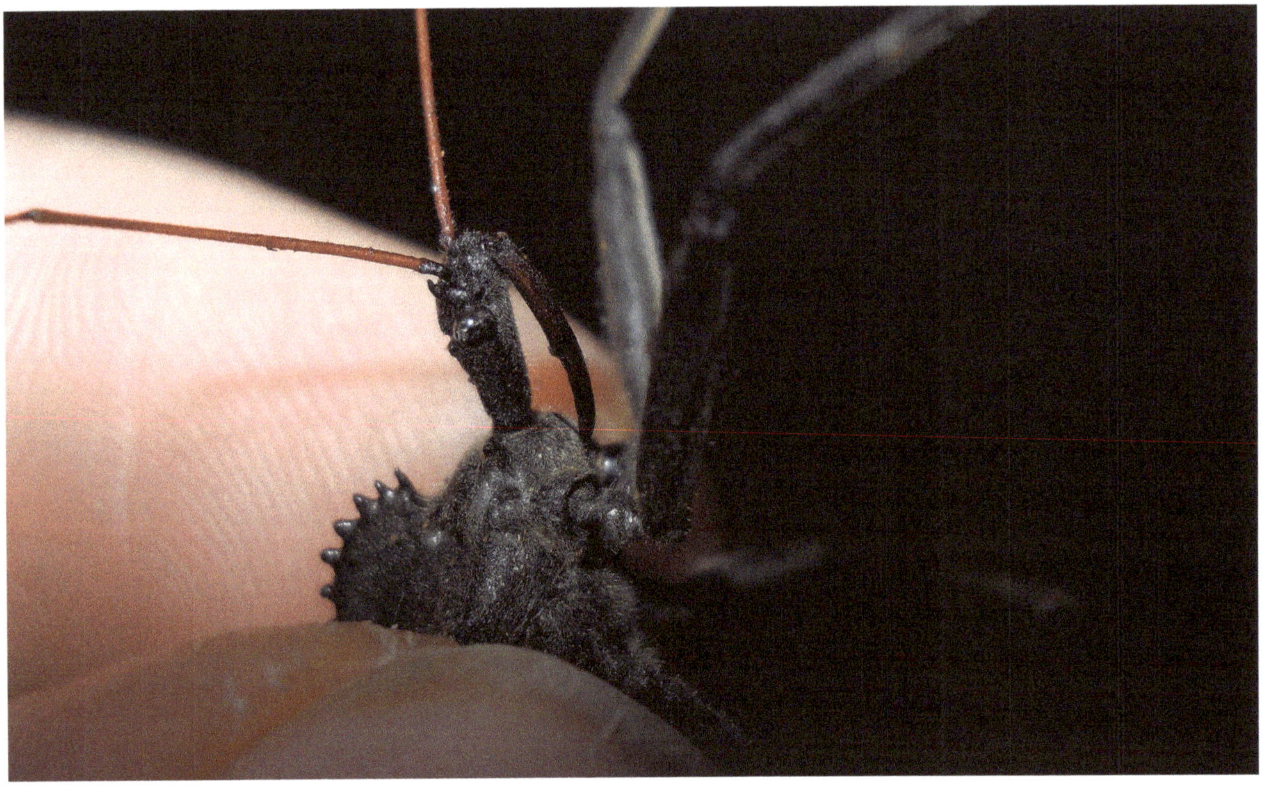

Arilus cristatus structure on the pronotum resembles half a cogwheel, specimen from Berlin, Ohio.

Only a small number of blood-feeding species are able to transmit vertebrate disease, and then generally only within populations of the host species. Triatominae are mostly restricted to one or a small number of specific animal hosts. In the United States, according to the Centers for Disease Control, Atlanta, U.S., "... only rare vector-borne cases of Chagas disease have been documented." There has never been a confirmed case of a conenose bug of any species transmitting the protozoan in the United States (Stoops and Martin 1995). Even if an infected conenose of the right type were accidentally carried home from South America in luggage, and simply bit a person, the possibility of infection from the bite is zero. The disease is spread by the feces being rubbed into the eyes, mouth, or open sores. The greatest risk associated with Chagas disease is if it is left untreated and becomes chronic. 20-30% of those infected will develop serious medical problems over the course of their lives as a result of the infection. The other 70-80% are asymptomatic (per the CDC).

SELECT REDUVIIDS

Wheel Bug

Arilus cristatus (Linnaeus, 1763)

This is North America's most massive assassin bug; adults are usually around an inch and a quarter (2.8–3.3 cm), rarely 10% longer. There are an additional five species found from Mexico to South America: *A. auctus*, *A. carinatus*, *A. depressicollis*, *A. gallus*, and *A. nigriceps* (Moldonado 1990). *Arilus cristatus* is common in eastern and central North America and can be encountered in very cold temperate zones down to -20° F (-29° C). Despite apparent cold tolerance, specimens are much more commonly encountered in south-central and southeastern states. They overwinter as eggs so the adults die before the end of the year even when brought indoors. However, specimens reared from immatures in captivity live a year or more.

Arilus cristatus are relatively communal compared to other commonly kept assassin bugs. They do not seem to cooperate in bringing down prey, but wild nymphs can be found congregating on plants in groups of three to six animals during the first few instars (pers. obs). I have reared groups of 3-12 immatures in tiny 16 oz (950 ml) containers without a single case of cannibalism. Adult specimens do not attack each other.

Wheel bugs like other assassins, generally do not recognize flesh as food.

Despite the large size of adults, the time from first instar to adult is only about 100 days. Good ventilation is needed to successfully maintain these arboreal animals. They can be misted once a day and will drink droplets if the ambient humidity results in more water loss than fluids received from the prey. Nymph and adult stages are incredibly easy to care for, with similar requirements to mantids, except they do not cannibalize. Ova collected outdoors have

Wheel bug (*Arilus cristatus*). (cc) Sam Droege / USGS Native Bee Inventory and Monitoring Program

Spider assassin (*Stenolemus* sp.), Australia. (cc) Jean and Fred Hort

an excellent hatch rate unless parasitized, but those laid in the laboratory nearly always die. Too much moisture and they mold, too little and the eggs desiccate. I have tried everything I can think of, including basic incubators for arachnid and mantis eggs, but have not been able to determine a good methodology for hatching captive-laid *A. cristatus* ova.

Like most Harpactorinae, *Arilus* are arboreal and found a few feet above the ground on herbaceous plants. They are found in gardens, fields, and forest edge habitats. *Arilus cristatus* are considered beneficial because they feed on Japanese beetles and other crop pests. Organic gardeners are thrilled to see this species. Although diurnal, adult wheel bugs are attracted to lights and can be found on the walls of gas stations at night, feasting on other insects attracted to the lights.

Small nymphs are warningly colored black with an orange abdomen, but they become cryptic as they mature. Adult wheel bugs are able to employ nearly every defensive strategy in the Reduviidae toolbox. They can stridulate by moving the tip of the rostrum against a ridge located on the body between the front legs. The shrill chirping noise is designed to startle predators, but the wings can also be fluttered noisily or allow escape through flight. If noises fail they switch to color and stench. The animals have a bright-orange stink gland that can be everted from the end of the abdomen. The smell is strong and unpleasant by any measure. At the same time they can inflict a painful bite from the other end of the body (the rostrum).

Giant Thread-Legged Assassin

Emesaya brevipennis (Say, 1832)

This is among the longest of all assassin bugs, in the inch and a half range (3.0–3.7 cm), but it is incredibly long and thin. Nevertheless, the expansive (diagonal legspan of 4″ or 10 cm) black-and-white striped middle and hind legs make this an incredibly impressive animal. This assassin displays arboreal behavior and often hangs upside-down or from the side of the cage like a praying mantis. The forelegs look very similar to those of stick mantids, but the flamboyant, prehensile antennae and the needle-like mouth give away its allegiance to the Reduviidae. The eggs are small, at 1/12″ (0.2 cm) in length by a mere fraction of that in width. The white operculum and small spines on the sides of each egg are not easily seen without a magnifying glass. Ova are individually glued to the lid and upper corners of the cage, usually one to four each evening. They

A fat, egg-laying female *Emesaya brevipennis* from near Berlin Lake in Ohio.

Emesaya brevipennis eggs glued to the microscreen lid of a container.

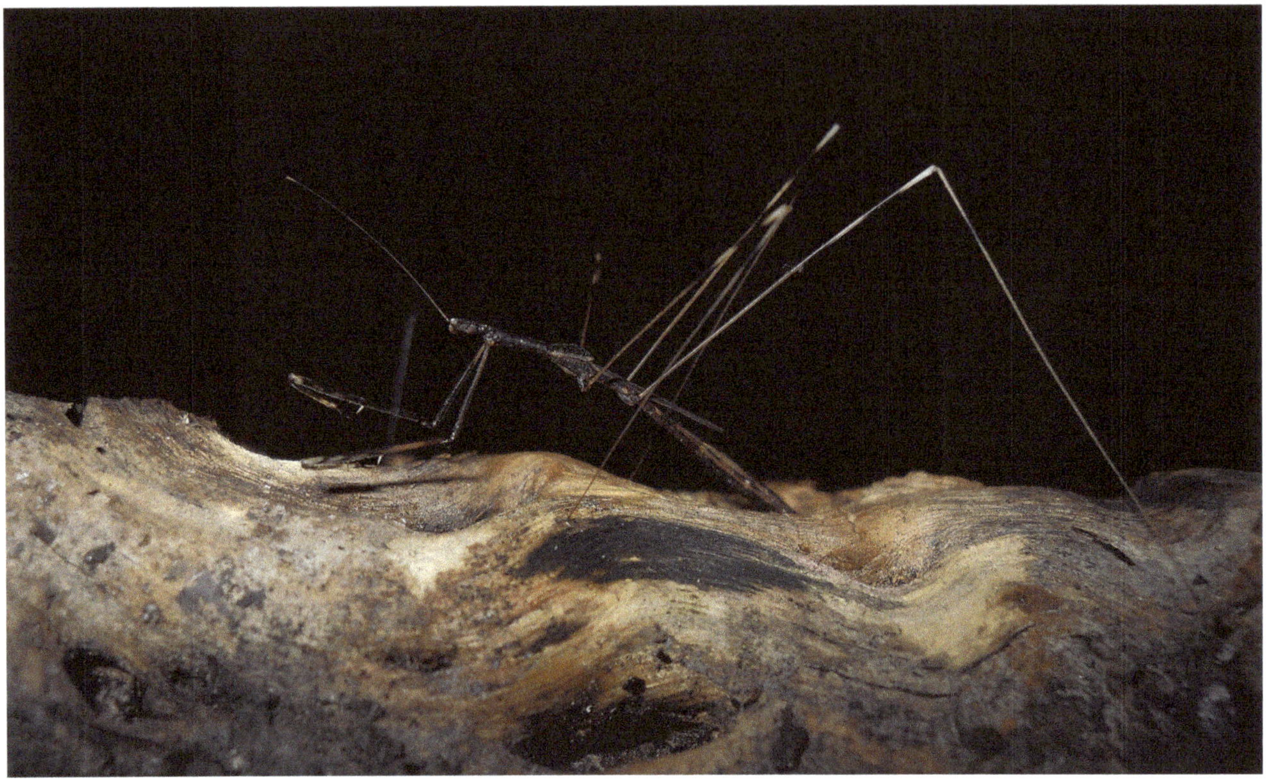

Emesaya thread-legged reduviid from Ohio.

The ocelli are easy to see in this image, male *M. picipes* black corsair from Alabama.

overwinter as eggs in nature but ova laid indoors hatch after about four weeks. Nymphs can take as little as two months to reach adulthood and the adults only live a few months afterwards. Immature specimens should be reared separately since they cannibalize in the later instars. I have reared this species from egg to adult on phorid flies but fruit flies can be used. Nymphs are prone to drying out but ventilation is required. Adults are apparently difficult to feed well which may be because their natural prey consists of web spiders, including orb weavers. The correct type of spiders may be difficult to acquire and likely offer a clear and present danger in captive situations.

Emesaya brevipennis hatchlings are so thin they are almost invisible.

Caging should include a screen surface lid—each can be reared in a 32 oz (950 ml) tall deli-cup with a 1″ (2.5 cm) square vent covered in microscreen. An enclosure (3.5″ / 9 cm diameter, 6″ / 15 cm high) or larger is needed for the last few molts because of the rather long legspan. The enclosure should be misted daily and substrate may be necessary if the ambient humidity is less than 25%.

Specimens are difficult to find in the northeastern U.S. during most of the year, but in late October to early November (after the first couple hard frosts) adults can be collected from under the awnings of buildings and sheds. They are often 9 ft. (approx. 3 m) above the ground and can be knocked down with a stick. It is amazing how hard it can be at times to find a specimen after it falls to the ground.

Bark Assassin Bug
Microtomus purcis (Drury, 1782)

This species is the only common *Microtomus* in the U.S., though *M. tibialis* is found in Texas (it looks nearly identical, but the legs are black). There are a total of 13 species (Moldonado 1990) including *M. cinctipes*, *M. conspicillaris*, *M. gayi*, *M kuntzeni*, *M. luctuosus*, *M. lunifer*, *M. maculatus*, *M. pessoai*, *M. pintoi*, *M. reuteri*, and *M. sticheli*.

Microtomus purcis from Florida kept in the late 1990s.

Eggs are laid individually, usually 1-4 each evening by females. Nymphs hatch after six weeks and can be fed *Drosophila*, followed after a few molts by small cockroach nymphs (for example *Schultesia lampyridiformis* or *Parcoblatta* spp.) and isopods. Large prey is not taken. Individual vials or cups are suggested for rearing immatures because they climb smooth vertical glass or plastic surfaces and readily escape through small holes.

Bark assassin nymphs should be kept in separate cages. *M. purcis*, Alabama stock.

White spot assassins, *P. biguttatus*.

Immatures are cannibalistic and cannot be kept safely in groups. Individual 5 oz (150 ml) to 16 oz (450 ml) plastic cups with vented lids and a piece of angled bark should be set up. A layer of paper towel, shallow sand, or dirt substrate can be used but is not required. They are more resistant to drying than one might imagine for a creature living under bark. Too much moisture can kill them. *Microtomus* are slow growing, but mature about a month quicker than *Platymeris* spp. Nymphs can reach adulthood in as little as 100 days, but more often take 150 days. Adults will have pink (instead of white) markings on the hemelytra for a number of hours following the ultimate molt.

This is a relatively long-lived species with a natural lifespan of two years or so. Many temperate assassins are naturally seasonal and the approximate month can be estimated by the instar. However, specimens of *M. purcis* ranging in age from second instar to adult have been found overwintering beneath bark (Jeon and Murphy 2015), suggesting seasons do not play a major part in reproduction in this species.

White Spot Assassin Bug

Platymeris biguttatus (Linnaeus 1767)

Platymeris comes from the Greek *platy* = "flat" and *meris* = "segmented." There are thirteen described species according to Maldonado (1990): *P. biguttata*, *P. charon*, *P. erebus*, *P. flavipes*, *P. guttatipennis*, *P. insignis*, *P. kavirondo*, *P. laevicollis*, *P. nigripes*, *P. pyrrhula*, *P. rhadamanthus*, *P. rufipes*, and *P. swirei*. The species name *biguttatus* means "two spots," but this could describe any *Platymeris* and many other unrelated assassins, or any other insect with two spots on the back. Most texts have the species name spelled *biguttata* instead of *biguttatus* because it had been the correct spelling of the species up until its redescription (Li, et al. 2010). This redescription was based strictly on laboratory stock since, despite its common presence in the laboratory for decades, there existed no detailed taxonomic description.

In addition to yellow bands on the legs and white spots on the wings, the upper surface of the abdomen (normally hidden by the wings) is red. Beyond color and size, adults can be differentiated from *P. rhadamanthus* and *Platymeris* sp. "Mombo" by the presence of extensive fine wrinkling on the surface of the pronotum. Fine wrinkling is minimal on *P.* sp. "Mombo" and is better described as shallow pits, while both pits and wrinkles are absent on the *P. rhadamanthus* adult's pronotum.

Wild animals are found under logs where they feed on ground-dwelling invertebrates. This and a few related species are found across forested areas of Africa. Although they can survive dry periods, they are not arid species.

The venom causes extreme pain. There are no records of allergic reactions to the bite, but airborne venom may cause allergic reactions in some people (this is difficult to properly isolate from allergies to cricket and cockroach prey). *Platymeris biguttatus* are capable of spitting venom but rarely more than an inch (< 2.5 cm). Some reports suggest venom can be spit as far as a foot (30 cm) (Larsen 1998) but such an event would be exceptional for this species. When I pick up adults they regularly spit on my fingers, but one would have to have them quite close to the eyes to be at risk. On the other hand, on multiple occasions, when keeping the *Platymeris* sp. "Mombo," the venom was spat in my eyes from at least 30 cm. The only reason this happened half a dozen or more times over a decade was because it was not a regular occurrence and discomfort was minimal. I rinsed out my eyes with warm water right away. It stung a little, but faded rapidly with prompt and proper care. (Waiting, rubbing the eyes, or rinsing with cold water would have compounded rather than alleviated discomfort.)

These large African assassin bugs are a common display animal since they are colorful, big, and easy to care for. Adults are normally 3.5–3.8 cm and can be kept safely in huge groups. They feed on crickets, but will also eat cockroaches, beetles, isopods, mealworms, and other invertebrates. Rearing cages for immatures should contain paper egg carton rather than decorative bark since the increased area greatly improves survival. Once the colony reaches a good size and a high survival rate is no longer reasonably maintained, nymphs can be left with adults.

Adults look similar from above but from below the genders are easy to discern. The male's abdomen ends in a cup-like segment covering the genitalia. As they are kept in colonies, fertility is not an issue as long as both genders are present. Females lay small brown to black eggs with a bright white operculum (or lid). These can be buried in shallow pits in damp areas of the cage or dropped below the adult's hide spot. Ova should be kept damp, but will survive limited dry periods and can even recover from temporary dimpling (where the egg shell contracts inwards from loss of moisture). Colonies reproduce well with all stages kept and fed together but initial startups can take some time, space, and effort due to cannibalism.

Red Spot Assassin Bug

Platymeris rhadamanthus Gerstaecker, 1873

Rhadamanthus was a judge of the underworld in Greek mythology. The stock kept round the world and depicted in the photographs in this text has been kept in continuous culture since the mid-1990s (possibly the 1950s). There has been some confusion with the scientific name. This species is often labeled *Platymeris laevicollis* (*laevicollis* comes from the Latin root words for "smooth" and "mound or hill," probably referring to the smooth pronotum). It has been reared somewhat regularly in laboratories at least as far back as the mid-1950s for release to control adults of the pestiferous coconut palm beetle, *Oryctes rhinoceros* (Vanderplank 1958, Man-januth et al. 1969, Sankaran 1974). This may be surprising to those familiar with the beetle because *O. rhinoceros* is heavily armored and many times the mass of the largest *Platymeris*.

Egg to adult *P. rhadamanthus*.

Mombo, once the most impressive and largest of the laboratory assassin bugs.

Psytalla horrida mixed stages group, subadults have notable wing pads. © Peter Grabowitz

The familiar stock of this species averages about 20% less massive than the white spot, usually 3.2 cm–3.6 cm, is less robust, and apt to ruffle its wings when disturbed. Specimens are not aggressive towards keepers and do not bite if touched or held, but most certainly bite when grabbed or restrained. They spray droplets of venom from the rostrum when greatly disturbed and this produces a rather potent odor. The odor is a musty, fruity smell that might be pleasant to someone who has not sustained a good bite. (Those who have been recently bitten will find the smell nauseating.) In this species the droplets are rarely seen and unlikely to get into the eye unless the animal is an inch (2.5 cm) from the eyeball.

Platymeris eggs usually hatch after four or five weeks but dry periods have an uncertain effect on hatch time. Vanderplank (1958) tested *P. rhadamanthus* eggs by keeping them dry for 96 days and the eggs hatched 20 days after contact with moisture. After hatching, 180 days or more are required for nymphs to reach adulthood. Prey availability will impact the length of time needed.

This species has been hybridized in captivity with *P. biguttatus*. Early attempts at isolating virgin adults and pairing them resulted in infertile eggs that did not hatch. "I had isolated subadult females and attempted numerous times (male-female and female-male) to cross white spots and red spots over the years without a single fertile egg" (McMonigle 2003). However, hybridization appears to have been achieved by a much less time consuming and less exact method. Haphazardly mixing multiple pairs together and maintaining them as a single colony worked where controlled experimentation and individual isolation failed. It should go without saying that different *Platymeris* species should not be kept in the same enclosures to prevent hybridization.

Orange Spot Assassin Bug
Platymeris sp. "Mombo"
This is a gigantic African species and for a decade it was the single largest assassin bug propagated in captivity. (It averages a little larger than the other two *Platymeris* stocks at 3.7–4.1 cm.) The original adults were collected under tree bark in the small town of Mombo in the southern foothills of the Usambara Mountains, Tanzania, around 1999 (Hromadka 2003). In the last few years, similarly colored stock from Nigeria or Cameroon, and identified as *P. rhadamanthus*, has shown up in European breeding (pers. comm. Schwarz). A properly fed female can grow to one and a half times the mass of *P. biguttatus* or *P. rhadamanthus*. (Similar specimens, according to Chlond et al. (2014), and the standard red spots are simply two different forms of *P. rhadamanthus*.)

Orange spots can gather in appropriate hiding spots, but the behavior is nothing like the clustering seen in false assassins such as *Euthyrhynchus*.

Adult coloration was originally somewhat variable in the first half dozen or so generations but today's stock tends to be very consistent. Nymphs still have quite a bit of variation as they grow, but the variation is standard within development time, not among individuals.

Platymeris and *Psytalla* size comparison (both are average size adult males).

The original 2009 wild-caught mother *Psytalla* with F1 offspring. © Peter Grabowitz.

Psytalla `road warrior` adult.

Nymphal leg banding ranges from deep red, to orange-red, and hot pink as they grow and the abdomen ranges from red to black. Most adults have orange spots on the wings though the spots range in hue from yellowish orange to bloody orange depending on the food nymphs were raised on. The original stock included a recessive color morph in which the adults looked like a very large *P. biguttatus* with tan rather than yellow leg banding. Nymphs of this recessive morph were gray and white and looked nothing like nymphs of the dominant form. (For that matter they looked nothing like the colorful nymphs of the other two species.) However, as a recessive form it was easily lost in captivity. It is far easier to remove or select for a recessive trait than to keep it in balance. The gray form was considered undesirable and has not been reported in cultures since circa 2005 (photos in this text are 2004 and earlier).

Platymeris (all species) are voracious hunters that seem to prefer oversized prey. These are the only invertebrate predators that will eat adult Madagascan hissing cockroaches (*Gromphadorhina portentosa*) regularly. (Some of the largest tarantulas, centipedes, or mantids may take out an adult hisser on the rare occasion if starved.) A small *Platymeris* colony can be used for continuous culling of color defects or small adults in hisser colonies.

Platymeris sp. "Mombo" white form adult in 2002.

White "Mombo" nymphs (right) appear to have been lost from culture around 2004 (photo 2003). Nymphs feeding on a cricket; look closely for the third, smaller nymph sharing in the meal.

Horrid King Assassin Bug

Psytalla horrida (Stal, 1865)

There was some confusion with this species since it originally showed up as "*Platymeris*" *horrida* on stock lists. There is no *horrida* in the *Platymeris* genus, though *Psytalla* was considered a subgenus of *Platymeris* for quite some time. (Stal's type specimen deposited in the Swedish Museum of Natural History is listed under *Psytalla* but labeled "Platymerus horrida.") All stock of *P. horrida* appears to have originated from a single female imported to Germany from Cameroon in 2009 (pers. comm., P. Grabowitz). The horrid king assassin bug is widespread in tropical Africa including Cameroon, Congo, Gabon, Guinea, Ivory Coast, Kenya, Malawi, Mozambique, Nigeria, and Central African Republic. There are five other *Psytalla* species: *P. ducalis*, *P. dudgeoni*, *P. incognita*, *P. johnstoni*, and *P. samwelli* (Maldonado 1990).

Psytalla nymphs seem relatively tolerant of conspecifics (teneral 4th instar nymph at center).

Psytalla horrida fresh subadult.

Freshly molted *Psytalla horrida,* 47 mm female.

1st and 2nd instar *P. horrida* nymphs.

According to McGavin's (1993) magnum opus on true bugs, the largest Reduviidae reach only 4.0 cm. This species measures 4.4 to 4.8 cm, nearly two inches long. The horrid king assassin bug is arguably the most spectacular and easily the most massive assassin bug ever maintained in captive culture. Those who have propagated *Platymeris* are already familiar with the requirements for *Psytalla*: moderate moisture, rough molting surfaces, and large prey. The only difference is this assassin outgrows small food more rapidly simply because each instar is one size ahead (even large fruit flies and pea weevils can be too small of prey for hatchling *Platymeris* and are far too small for these). A 1st instar *P. horrida* is the size of a 2nd instar *P. rhadamanthus* and accordingly should be fed larger prey. Subadults readily kill adult hissing cockroaches. The first four instars take approximately five weeks each, while the fifth and final instar is nine weeks or more. From hatching, specimens can take as little as 210 days to mature; on average a month longer than *Platymeris*. Some specimens in any group will grow more slowly by 30-60 days. Reduced or inadequate feeding can easily double the required time for an animal to reach maturity.

Zelus longipes nymphs are amazingly active predators that go into a feeding frenzy when offered fruit flies or pea weavils. This species normally does not cannibalize and is very easy to rear in captivity. The specimens shown here are from Lafayette, Louisiana, but *Z. longipes* is also common in Central and South America.

Like *Platymeris* and *Microtomus*, these are nocturnal but will feed any time of day in captivity. They hide under cage decorations during the day and look for large spaces (versus tight cracks favored by some bugs), which can be provided easily with paper egg carton, or paper carton packing for electronics.

When an adult *Psytalla horrida* specimen is picked up by the pronotum, it stridulates notably. The rostrum is rubbed against a row of small ridges between the front legs to make a chirping, rasping sound. *Platymeris* has the same structures for stridulation, but I have not heard a single specimen of any species use this structure in two decades. I have only heard the common wing buzzing of *P. rhadamanthus*. The second part of the *P. horrida* defensive response is the same as *Platymeris*: projectile venom. Although spitting of the venom was not directly observed, the fingers that held the pronotum had an undeniably strong smell of venom. The odor is not eliminated with one good hand washing. The bite is likely very similar to that of *Platymeris* because the odor of the venom and venom quantity appear similar.

"This thread-legged assassin, *Ischnobaena* sp., is found at the base of larger trees on mossy bark. Spiders may be regular or occasional prey for this species." Panay rainforest. Pers. comm. and © Christian Schwarz. Emesinae commonly use their antennae like a pair of front legs for balance. The front legs are reserved for prey capture and manipulation.

An arboreal and nocturnal *Canthesancus* sp. in Panay. © Christian Schwarz

"Small Emesinae with long wings usually found under leaves in the forest." Panay Island, Philippines. Pers. comm. and © Christian Schwarz

"A typical harpactorine, found on leaves in sunlit places and small clearings inside the forest." Panay Island, Philippines. Pers. comm. and © Christian Schwarz

Sinea sp. nymph, Cincinnati, Ohio. © Martin Gamache

Pristhesancus sp. sporting a trio of strange knobs on the pronotum of the adult, Panay. © Christian Schwarz

Camouflaged assassin bug nymph, *Lisarda* sp., Panay Island, Philippines. "The nymphs of this species live under a research station building and in other dry places in the rainforest. The adults are also found in the rainforest. They seem to feed mainly on ants and termites in nature. Walking dust bunnies like this one are not necessarily Reduviinae, several Harpactorinae also practice this camouflage." Pers. comm. and © Christian Schwarz

A nocturnal assassin bug, *Lisarda* sp., feeding on its usual prey, a winged ant, Panay. Nymphs of this species are camouflaged with debris. © Christian Schwarz

“A terrestrial assassin bug, sometimes found on rotten logs or low vegetation, but mostly on leaf litter. It tends to aggregate to some degree, at least there are usually a few where there is one.” Panay. Pers. comm and © Christian Schwarz

Unidentified assassin nymphs in the rainforest, Panay. © Christian Schwarz

An apterous *Ischnobaena* sp. (Emesinae) feeding on a spider, Panay Island, Philippines. © Christian Schwarz

The smaller male *Phymata* sp. is commonly found guarding the female. He sits on top but is rarely seen mating, Montreal, Canada. © Martin Gamache

Phymata sp. are able to change color to match the flowers they hunt on. The change takes place after molting so the color of this animal is fixed. Adult female, Montreal. © Martin Gamache

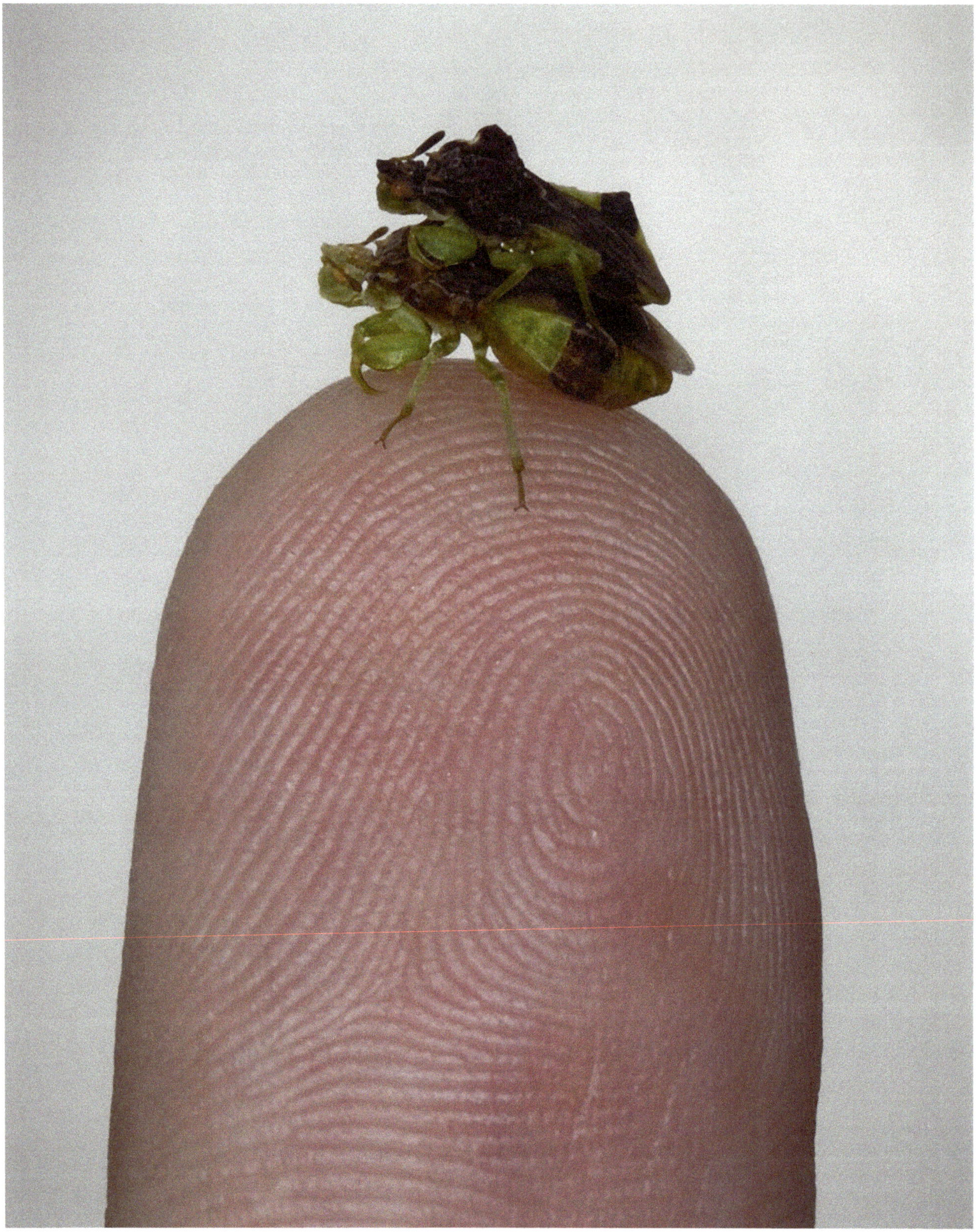

Phymatinae are very small bugs. © Henry Kohler

EPILOGUE

TRUE BUGS ARE HARDY and spectacular animals for use in laboratory observations and invertebrate displays. Their greatest value is in living, breathing creatures and not just as dried curios or museum drawer specimens. Continued exploration for their use as agents of biological pest control and producers of chemicals for use in future industry and medicine is certainly worthwhile. Preservation of captive stocks ought to become paramount as humans continue to eliminate natural ecosystems. Heteropterans' value as living creatures to be studied, displayed, and enjoyed should not be overlooked. The husbandry information provided here is meant to offer a starting point for continued work with these spectacular creatures. Hopefully the photos and information presented here will ignite the passion of those considering keeping these incredible creatures. Investigation of our planet's incredible fauna, including bugs, deserves consideration. If only it would rank in value with even the less popular sports or video games.

Phymatinae male guarding the female. © Henry Kohler

This assassin bug nymph is holding a large ball of *Nasutitermes* carcasses. It uses the dead termite bodies to lure more termite prey, French Guiana. © Martin Gamache

Wheel bug feeding on captured bee. (cc) Photochem_PA

GLOSSARY

ABDOMEN: The terminal section of the insect body containing the gut and genital organs.

APTEROUS: Without wings.

COXA (*pl.* COXAE): Basal segment of the insect leg connecting the body to the trochanter.

CRYPSIS: Homochromy or background-matching; mimicry of common background objects like rocks or plants rather than fauna.

DESICCATION: Drying out, often to the point of death.

DIAPAUSE: A period of inactivity or arrested development influenced by seasonal changes in temperature or humidity, such as overwintering.

DIURNAL: Day active, the opposite of nocturnal.

EVERSIBLE: Able to be turned inside-out.

EXOSKELETON: The external body covering of arthropods which serves as protection and muscular support. Mainly consisting of chitin and proteins.

FEMUR (*pl.* FEMORA): Section of the insect leg located between the trochanter and tibia; usually the largest segment of the leg.

FRASS: The semi-solid waste of bugs and other invertebrates.

GENITAL OPERCULUM: Modified male segment located towards the rear of the abdomen which covers the genitalia; specialized or secondary subgenital plate. For bug families such as Coreidae and Reduviidae, this segment allows for quick and accurate identification of the male.

GREGARIOUS: Living in groups or preferring to be near others of one's own kind.

HEMELYTRA: Outer, thickened wings that are membranous towards the ends.

HEMIMETABOLAN: Insects with incomplete metamorphosis.

HEMIMETABOLOUS: Type of insect life cycle composed of only three stages: egg, nymph, and adult (lacking a pupa stage); incomplete metamorphosis.

HULL: The dry outer covering of a seed, fruit, or nut.

HYDROPHOBIC: Property of being repellent to water.

INSTAR: Developmental stage between each molt and prior to maturity.

LABIUM: Tube-like outer covering of the rostrum.

MARGIN: A border or edge.

MEDIUM (*pl.* MEDIA): The food as well as substrate used in culturing fruit flies, bacteria, etc.

MOLT: *verb*, to shed the exoskeleton or skin. *noun*, the act of shedding or replacing the outer shell or skin; the discarded skin/shell.

NYMPH: The premature stages of insects with a hemimetabolous life cycle.

Reduvius personatus adult female, Lafayette, Louisiana.

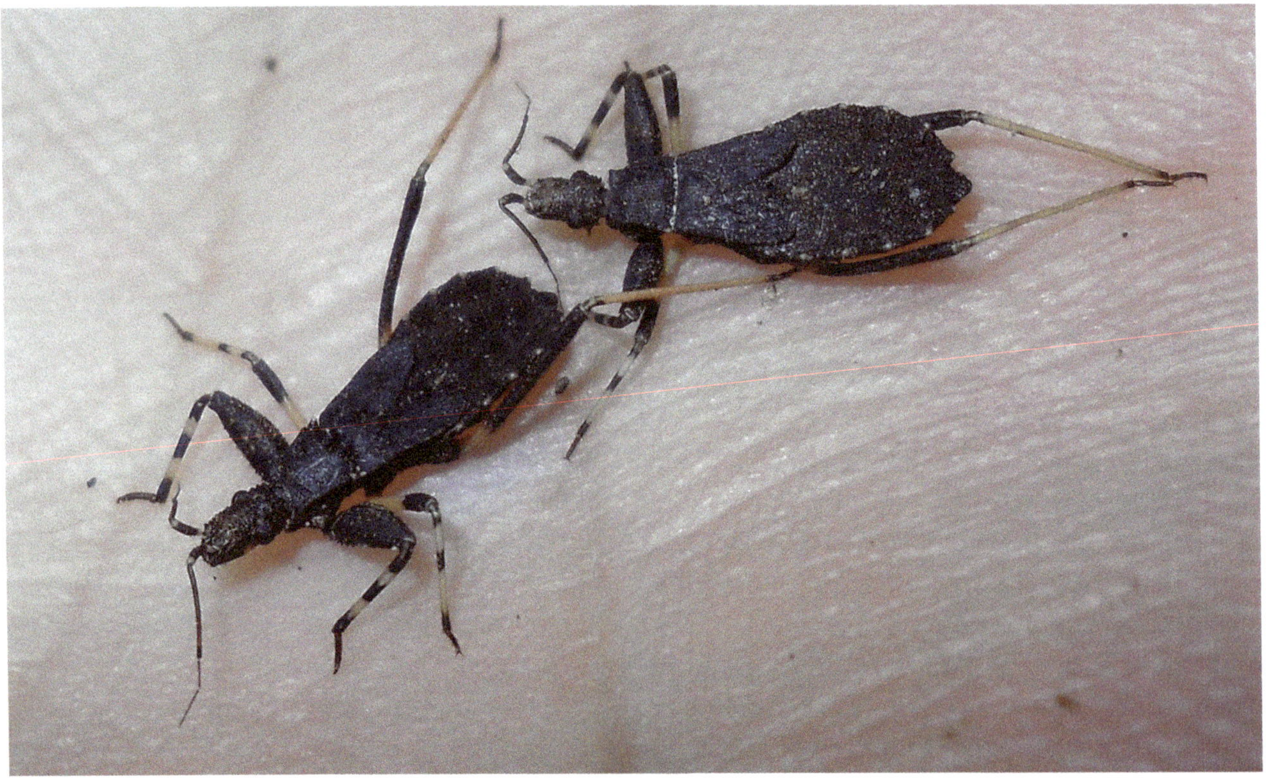

Subadult specimens from outside Alexandria, Louisiana. Found under rocks. They play dead during daylight hours. Members of the subfamily Stenopodainae such as *Stenopoda* sp. and this unidentified species, have proven difficult to feed in captivity.

OCELLUS (*pl.* OCELLI): Light detection organs. Often arranged in a set of three on top of an insect's head between the compound eyes.

OPERCULUM: Specialized section of the bug egg through which hatchlings emerge, resembling a trapdoor. This term is used to describe various lid-like or door-like coverings on plants and animals. *See also* genital operculum.

OVIPOSIT: To lay or deposit eggs.

OVUM (*pl.* OVA): Egg.

PALUDARIUM: A terrarium that incorporates terrestrial and aquatic zones.

PARTHENOGENESIS: Reproduction through the development of unfertilized eggs.

PRONOTUM: Shield-like upper surface of prothorax.

PROTHORAX: Front section of the thorax; attached to the head and contains front pair of legs.

PULVILLUS (*pl.* PULVILLI): Adhesive, fleshy pad located between the tarsal claws which allows a bug to climb smooth surfaces such as glass.

RAPTORIAL: Having a curved shape useful in capturing prey, designed for grasping like the claws of birds of prey.

ROSTRUM: The distinctly designed sucking mouth of hemipterans, composed of several segments and variably adapted for different foods.

SCUTELLUM: The rear portion of the mesothorax, originates on the dorsal surface, in-between the wings; usually triangular in shape although sometimes greatly enlarged and rounded.

SETA (*pl.* SETAE): Hollow growth from the epidermal layer of the body wall: hair-like setae are often referred to as "hair" but are principally different from the hair of mammals.

SEXUALLY DIMORPHIC: Presence of noticeably different external gender characteristics.

STRIDULATE: To create noise by rubbing together body parts, usually specialized legs or wings.

STYLET: Needle-like central portion of the rostrum; includes the maxillae and mandibles.

SUBADULT: Nymphal stage immediately preceding maturity, 5th instar in true bugs.

SUBGENITAL PLATE: Segment covering the ventral surface of the genitalia; genital operculum.

TENERAL: Soft, often pale or white colored state of the insect exoskeleton immediately following a molt.

TARSUS (*pl.* Tarsi): The foot of an insect, made up of a number of small segments including the pulvilli and claws.

THANATOSIS: The act of feigning death in order to avoid attention.

TIBIA (*pl.* TIBIAE): Segment of an insect's leg located between the foot (tarsus) and the femur.

TROCHANTER: Segment of the insect leg, located between the coxa and femur.

TRUNCATE: Short, cut off, or appearing to be cut off at the tip.

ULTIMATE MOLT: The final shedding of the exoskeleton from an immature and resulting in a sexually mature specimen.

VECTOR: A species that transmits a certain disease.

WINGPAD: Undeveloped wings of nymphs, which appear as flat structures.

Damselbugs in the family Nabidae are predatory and resemble small reduviids but have a four-segmented rostrum. Montreal, Canada. © Martin Gamache.

Reduvius personatus nymph, Montreal Canada. © Martin Gamache

BIBLIOGRAPHY

Arnett, Ross H. Jr. (1993, 2000) *American Insects: A Handbook of the Insects of America North of Mexico*. Sandhill Crane Press / CRC Press.

Barclay, Maxwell. (1996) Bug unmasked. *The Bulletin of the Amateur Entomologists' Society* 55(404): 45-46.

Brown, Lara N., and J. E. McPherson. (1994) Life history and laboratory rearing of *Gelastocoris oculatus oculatus* (Fabricius) (Hemiptera: Gelastocoridae) with descriptions of immature stages. *Proceedings of the Entomological Society of Washington* 96(3): 516-526.

Bruins, Eugene. (1999) *Encyclopedia of Terrarium*. Rebo Productions, Lisse, Netherlands.

Cassis, G., and G. F. Gross. (2002) *Zoological catalogue of Australia: Hemiptera: Heteroptera (Pentatomomorpha)*. CSIRO Publishing, Clayton, Victoria, Australia.

Chlond, Dominik. (2010) A new, remarkable genus and two new species of Ectrichodiinae (Hemiptera: Heteroptera: Reduviidae) from Madagascar. *Zootaxa* 2522: 61-68.

Chlond, Dominik, and Agnieszka Bugaj-Nawrocka. (2014) Model of potential distribution of *Platymeris rhadamanthus* Gerstaecker, 1873 with redescription of species. *Zoological Studies* 53(8). (online) doi:10.1186/1810-522X-53-8

Cranshaw, W. S., M. Camper, and F. B. Peairs. (1997-2013) *Bat Bugs, Bed Bugs and Relatives*. Colorado State University Extension. www.ext.colostate.edu/pubs/insect/05574.html

Degeer, Charles. (1773) *Memoires pour Servir à l' Histoire des Insectes*. Pierre Hesselberg, Stockholm.

Dunn, Gary A. (1993) *Caring for Insect Livestock: An Insect Rearing Manual*. Young Entomologist's Society, Lansing, MI.

Eaton, Eric R., and Ken Kaufman. (2007) *Kaufman Field Guide to Insects of North America*. Houghton Mifflin, New York, NY.

Edwards, John S. (1961) The action and composition of the saliva of an assassin bug *Platymeris rhadamanthus* Gaerst. (Hemiptera, Reduviidae). *Journal of Experimental Biology* (38): 61-77.

Eluwa, M. C. (1972) The life history and egg laying behaviour of *Platymeris biguttata*. *Bulletin of the Entomological Society of Nigeria* (3): 108-120.

Forthman, Michael, and Christiane Weirauch. (2012) Toxic associations: A review of the predatory behaviors of millipede assassin bugs (Hemiptera: Reduviidae: Ectrichodiinae). *European Journal of Entomology* 109: 147-153.

Fracker, Stanley B. (1912) *A Systematic Outline of the Reduviidae of North America*.

Robert Henderson, Des Moines, IA. (rpnt. from *Iowa Academy of Science* 19: 217-252.)

Fritzsche, Ingo. (2003) Die zweifleck-raubwanze, *Platymeris biguttatus* (Linne, 1767) (Heteroptera; Reduviidae). *Arthropoda* 11(4): 2-7.

Hagerty, A. M., and J. E. McPherson. (2000) Life history and laboratory rearing of *Arilus cristatus* (Heteroptera: Reduviidae) in Southern Illinois. *Florida Entomologist* 83(1): 58-63.

Hagerty, A. M., J. E. McPherson, and J. D. Bradshaw. (2001) Life history and laboratory rearing of *Emesaya b. brevipennis* (Heteroptera: Reduviidae) in southern Illinois. *Florida Entomologist* 84(3): 410-414.

Hawkeswood, Trevor. (1992) *Pristhesancus plagipennis* Walker, a large tropical assassin-bug from Australia (Hemiptera: Reduviidae). *Bulletin of the Amateur Entomologists' Society* 51(380): 22-23.

Henry, Thomas J. (1997) Phylogenetic analysis of family groups within the infraorder Pentatomomorpha (Heteroptera), with emphasis on the Lygaeoidea. *Annals of the Entomological Society of America* 90(3): 275-301.

Henry, Thomas J. (2009) Biodiversity of the Heteroptera. in: Foottit, R. G., and P. H. Adler, eds. *Insect Biodiversity: Science And Society*. Wiley-Blackwell, Oxford. pp. 223-263.

Howarth, Francis G., and William P. Mull. (1992) *Hawaiian Insects and Their Kin*. University of Hawaii Press, Honolulu.

Hromadka, Jiri. (2003) Not only to see flower beetles in Usambar. *Sklipkan Journal for Invertebrates Keepers and Breeders* 8(3): 107-115.

Hutchins, Ross E. (1966) *Insects*. Prentice-Hall, Englewood Cliff, NJ.

Jeon, Alan, and Joe Murphy. (2015) Beneath the bark: *Microtomus purcis*. *Invertebrates-Magazine* 14(2).

Kleinsteuber, Erich. (1989) *Kleintiere im Terrarium. Wirbellose halten, züchten, kennenlernen*. Urania Leipzig, Germany.

Larsen, Tom. (1998) *Platymeris biguttatus*– om Rovtaeger! *Exotiske Insekter* (7): 10-12.

Li, H., et al. (2010) Taxonomic and bionomic notes on the white spot assassin bug *Platymeris biguttatus* (Linnaeus) (Hemiptera: Reduviidae: Reduviinae). *Zootaxa* 2644: 47-56.

Löser, Siegfried. (1991) *Exotische Insekten*, Tausendfüßer und Spinnentiere. Ulmer Verlag, Germany.

Maldonado, Capriles J. (1990) *Systematic Catalogue of the Reduviidae of the World (Insecta: Heteroptera). Caribbean Journal of Science*, Special publication No. 1, 694 pp.

Manjunath, T. M., M. K. Kamath, and V. P. Rao. (1969) Investigations on natural enemies of *Oryctes rhinoceros* (L.) (Col., Scarabaeidae) in India. *CIBC Technical Bulletin* (11): 65-91.

McGavin, George C. (1993) *Bugs of the World*. Blandford, London.

McMonigle, Orin. (2001a) *Assassins, Waterscorpions & Other True Bugs Care and Culture*. Elytra and Antenna, Brunswick, OH.

McMonigle, Orin. (2001b) The ferocious water bug, *Abedus herberti*. *Insecta* 1(1): 26-28.

McMonigle, Orin. (2003) *Orange spot assasin adult pic*. [sic] (8-12-2003) http://www.arachnoboards.com/ab/archive/index.php/t-13184.html

McMonigle, Orin. (2004a) Giant African waterscorpions. *Invertebrates-Magazine* 3(3): 5-9.

McMonigle, Orin. (2004b) The ferocious water bug. *International Fauna Society's The Keeper*. (June): 4.

Rasahus hamatus, orange spotted corsair. A) Female. B) Male. C) Ova. D) 1st instar nymphs. This is a small species that likes high humidity. Calera, Alabama. © Joe Murphy

McMonigle, Orin. (2011) *Invertebrates for Exhibition Insects, Arachnids, and Other Invertebrates Suitable for Display in Classrooms, Museums, and Insect Zoos.* Coachwhip Publications, Landisville, PA.

McMonigle, Orin. (2014) *Euthyrhychus floridanus* fantasmogoric false assassin. *Invertebrates-Magazine* 13(4): 17.

Morgan, Randy C. (1996) *Why Display Water Striders?* AZA Western Regional Conference. (proceedings)

Morgan, Randy. (1988) Windows on the water world. *Backyard Bug Watching* No. 5.

Morgan, Randy, and Ellen Thoms. (1994) Insects: diving, swimming, snorkeling, floating. Cincinnatti Zoo & Botanical Garden booklet article, pp. 6-9.

Murphy, Joe. (2014) An overview of the Florida predatory stink bug (*Euthyrhynchus Floridanus*). *Invertebrates-Magazine* 14(1): 10-12.

Perez-Goodwyn, Pablo J. (2006) Taxonomic revision of the subfamily Lethocerinae Lauck & Menke (Heteroptera: Belostomatidae). *Stuttgarter Beitr. Naturk.* 695(71): 74.

Poisson, Raymond-Alfred. (1965) Catalogue de Heteropteres Hydrocorises africano-malgaches de la famille des Nepidae (Latrielle) 1802. *Bulletin de 1 I.F.A.N.* T XXVII, Ser. A, No. 1.

Rees, A. R., and R. E. Offord. (1969) Studies on the protease and other enzymes from the venom of *Lethocerus cordofanus. Nature* 221(5181): 675-677.

Reisenman, C. E., et al. (2011) Feeding and defecation behavior of *Triatoma rubida* (Uhler, 1894) (Hemiptera: Reduviidae) under laboratory conditions, and its potential role as a vector of Chagas disease in Arizona, USA. *American Journal of Tropical Medicine and Hygiene* 85(4): 648-656.

Ribiero, J. R. I. (2007) A review of the species of *Belostoma* Latreille, 1807 (Hemiptera: Heteroptera: Belostomatidae) from the four southeastern Brazilian states. *Zootaxa* 1477: 1-70.

Sankaran, T. (1974) Natural enemies introduced in recent years for biological control of agricultural pests in India. *Indian Journal of Agricultural Sciences* 44(7): 425-433.

Stanek, V. J. (1969) *The Pictorial Encyclopedia of Insects.* Hamlyn House, London.

Stoops, E. D., and J. L. Martin. (1995) *Scorpions and Venomous Insects of the Southwest.* Golden West Publishers, Phoenix, Arizona.

Vanderplank, F. L. (1958) The assassin bug, *Platymerus rhadamanthus* Gerst (Hemiptera: Reduviidae), a useful predator of the rhinoceros beetles *Oryctes boas* (P.) and *Oryctes monoceros* (Oliv.) (Coleóptera: Scara-baeidae). *Journal of the Entomological Society of Southern Africa* 21(1): 309-314.

Wheeler, Alfred G. (2001) *Biology of the Plant Bugs (Hemiptera: Miridae): Pests, Predators, Opportunists.* Cornell University Press, 507 pp.

Wiley, Grace O. (1922) Some notes on the biology of *Curicta* from Texas. *The University of Kansas Science Bulletin* 14(20): 505-510.

Yelich, Alex. (2004) Killer bugs. *Invertebrates-Magazine* 3(2): 10-14.

"This *Eulyes* species, at around 3 cm, is the largest Harpactorinae on Panay Island. Like *Sycanus* and *Pristhesancus*, it is active by day and flies from leaf to leaf in sunlit places in the rainforest. *Eulyes* is rather rare on Panay, I only found two specimens in three years. I kept one alive to discover its food preferences. It was picky and would not touch anything except caterpillars, and those rather reluctantly. Caterpillars that were too large or too hairy were refused. Ants were refused as well. I do not know if caterpillars are its natural food. If this species was able to adapt to other commonly used feeders like flies, waxmoths, crickets, etc., breeding would still be complicated. It is questionable if they can be kept communally, in nature they are strictly solitary." Pers. comm and © Christian Schwarz

Assassin bug, Australia. (cc) Ian Sutton

Assassin bug nymph, Australia. (cc) James Niland

ABOUT THE AUTHOR AND EDITORIAL CONTRIBUTORS

Orin McMonigle has published half a dozen articles on captive husbandry and breeding heteropterans, and authored *Assassin, Water-scorpions and Other True Bugs* (2001).

Christian Schwarz is an entomologist and tropical ecologist who has worked on dozens of scientific texts as well as providing editorial assistance on arthropod books.

Annie Lastar has been assistant editor for *Invertebrates-Magazine* 2001-2015 and has edited a variety of other texts.

Cameron Campbell is a former zoo insectarium keeper of ten years, and is the author of the natural history website, naturalworlds.org.

Male *M. picipes* eating a cowpea weevil.

COACHWHIP PUBLICATIONS

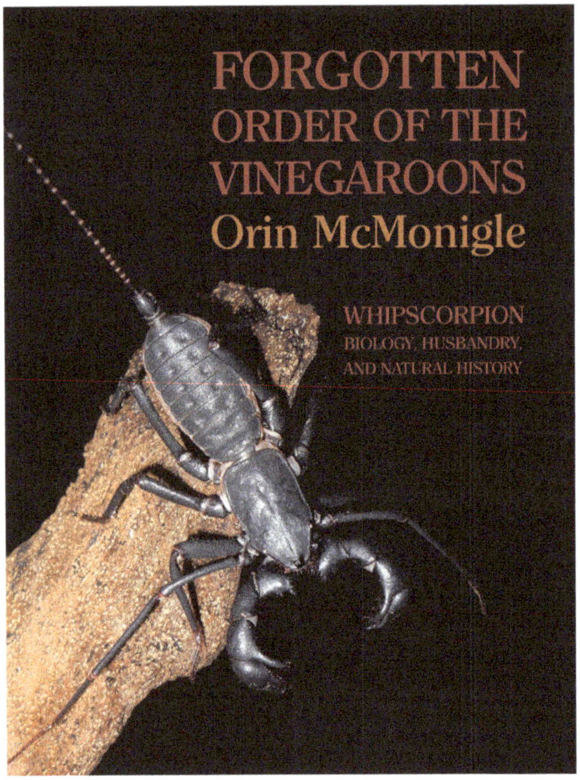

www.ingramcontent.com/pod-product-compliance
Ingram Content Group UK Ltd.
Pitfield, Milton Keynes, MK11 3LW, UK
UKHW061852121026
15625UKWH00066B/492

9 781616 464264